THE
HIDDEN
WORLD

**How Insects Sustain Life on Earth Today
and Will Shape Our Lives Tomorrow**

GEORGE McGAVIN

WELBECK

Published in 2023 by Welbeck
An imprint of Welbeck Non-Fiction Limited
Part of the Welbeck Publishing Group
Offices in: London – 20 Mortimer Street, London W1T 3JW &
Sydney – 205 Commonwealth Street, Surry Hills 2010
www.welbeckpublishing.com

A CIP catalogue record for this book is available from the British Library.

ISBN 978-1-80279-493-9

Typeset by seagulls.net
Printed in Great Britain by CPI Books, Chatham, Kent

10 9 8 7 6 5 4 3 2 1

CONTENTS

PREFACE

I'm a biologist and I have a particular passion for insects. Throughout my career, in both academia and broadcasting, I've learned about the species you might rather disparagingly call 'bugs' or 'creepy-crawlies'. We undervalue and denigrate these species at our peril. Insects are *amazing*. And they are essential to the normal functioning of our home, planet Earth.

In my lifetime, we have made enormous scientific and technological advances. The human population has risen from around 2.5 billion when I was born in the 1950s to nearly 8 billion now. The changes that have taken place on Earth during this time have been greater than at any other time in our history. I am now worried that we have lost sight of what really matters. That is to say, I'm worried about insects – because insects *really* matter. They are the most diverse and abundant group of animals that have ever lived. They made the world. They maintain the world. But they are, too rapidly, disappearing from the world.

In this book, I want to tell you all about them. I want to analyse their ingenious behaviour, examine the threats facing them and sing their praises in conversation with seven others – entomologists, professors and insect ambassadors who have spent their time, professional and personal, admiring these creatures small and great.

I want to tell you about how important they are, how interesting and strange they can be, and why I think it's high time all of us should care deeply about their fate.

Because if the creatures that made the world disappear entirely, I'm pretty sure that we will too.

Dr George McGavin
Ascot, 2022

Chapter 1

THE CREATURES FROM THE BLUE LAGOON

It's a small world

We live on a very small and insignificant rocky planet. The Sun, an average-sized star, around which we orbit, is 93 million miles away and light from it takes a little over eight minutes to reach us. Even at this huge distance, the Sun is visible because it is so large and contains most of the mass of our solar system. Less than a quarter of a per cent of the mass makes up all the planets, moons, asteroids and other stuff. Like a single particle inside a tonne of finely ground flour, our Sun itself is only one of many billions of stars in the galaxy that contains it.

We are mystified by extremes of magnitude, and find it very hard to visualise things that are not part of our everyday experience. Things we can see and touch and time frames that revolve around a day, a year or a human lifespan are easy to understand. But when we start thinking about extraordinary distances or durations, it gets difficult.

Lie on your back after dark anywhere the sky is not affected by light pollution and you will make out a bright, fuzzy band. I got my first spectacular view of our home galaxy, the Milky Way, when I was working in Tanzania, on a game reserve called Mkomazi. Lying on the warm ground and staring up at the black velvet heavens twinkling with innumerable glittering stars, I saw the broad, misty streak of the Milky Way arching across the darkness.

Humans have a fundamental need to understand and explain the complexities of the world around them. The ancient Greeks imagined that the Milky Way was milk that had been accidentally spilled when the goddess Hera was feeding the infant Heracles. In Hindu mythology it is regarded as a celestial version of the sacred River Ganges. The Milky Way appears as a band in the night sky because it is a flattened disc, of which our solar system is a very tiny part. Spinning at a couple of hundred kilometres per second, it is made up of somewhere between 100 billion and 400 billion stars, many of which may have a planetary system similar to ours. We now know that our solar system is located about halfway along one of the trailing arms that curves out from the centre of the Milky Way, like the streaks on the surface of a stirred latte. The diameter of the Milky Way is more than 200,000 light years, and at its centre is a massive back hole. Our planet is pretty inconsequential on a galactic scale and vanishingly so on a cosmic scale. Nevertheless, the Earth is our home and we share it with millions of other species.

My thoughts on that still African night were cut short by a loud rustling much closer to home. In an instant I was no longer concerned with the heavens above but rather with what might be about to attack me. The safety of the Land

Rover was tantalisingly close, but getting there might give my presence away. I lay absolutely still, wondering what to do. As it turned out, an aardwolf, a small species of hyena, was making its way through the undergrowth looking for food. Aardwolves are specialist insectivores with a particular penchant for termites, of which they may eat a quarter of a million in a single night of foraging.

Thinking about these very tiny creatures brought me back down to Earth. Termites are part of the most successful group of animals that has ever lived on Earth – the insects – and it is their epic story I want to tell. Humans overlook how tiny and inconsequential the Earth is compared to the vastness of space. But they also tend to overlook the importance of small creatures. Termites are integral to our world and insects conquered the planet long ago. They were among the very first animals to appear on land and were the first to take to the air. Their simple but versatile body plan has allowed them to colonise every habitat on Earth, and they are staggeringly numerous. Their total biomass is at least 10 times that of all humans and our livestock combined. It is hard to imagine a terrestrial or freshwater ecosystem that does not depend on insects. The world isn't ours at all. We are newcomers on a planet made and maintained by insects.

Six legs good

Insects belong to a huge group of animals called the Arthropoda, species with hard outsides and lots of pairs of hinged limbs. There are four distinct groups. There are the millipedes and centipedes with elongated bodies and lots of

3

pairs of legs, which you will find in soil and decaying wood, or under stones. There are the familiar spiders and other eight-legged species such as mites, ticks, harvestmen and scorpions. A third group comprises the crustaceans, such as shrimp, crabs, lobsters, crayfish, krill and barnacles. They are mainly marine but there are freshwater and a few terrestrial species, such as woodlice, as well. The fourth and by far the biggest group – and the most successful of all the arthropods – are the Insecta, which comprise three quarters of all animal species and well over half of all known species. It's worth taking a little time to let that fact sink in.

I have never understood how anyone could regard themselves as a zoologist without knowing about insects. Despite the overwhelming numerical superiority of insects, it is vertebrate species that attract most academic attention and conservation efforts. Insects that eat our crops and transmit diseases are feared and exterminated, while the vast majority are simply ignored.

I know that many people don't like insects that much and, while they might not find bees or butterflies too much to cope with, flies, beetles and wasps often get short shrift. So, when I give public talks, I often show a slide on which are the images of several different animals. There is a bumble bee, a hornet, a couple of flies, a beetle and even a head louse. But there is also a small image of a young bushbaby. The bushbaby looks dolefully at the audience with huge, wide eyes. Its tiny hands grip the fingers of the human hand that holds it aloft. The eyes of the audience are irresistibly drawn to the bushbaby as if the rest of the slide is completely blank. I have learned to pause for the inevitable and collective 'aww'. In all the years I have used this slide, I have never *not* heard the 'aww' sound from an audience and for this reason, I call the slide 'the aww factor'.

The sum total of all vertebrate species – fish, birds, amphibians, reptiles, mammals, the whole beastly bestiary, from aardvarks to zebras and bats to blue whales – makes up a little under 3% of all species alive today. It is difficult not to come to the conclusion that they are a pretty trivial group of animals compared to insects. Despite this, animals with a backbone really do seem to hold a very special place in the human heart. I suppose it's easy to see how a cute bushbaby would trump any number of insects. After all, bushbabies are primates just like us. They have large brains, forward-facing eyes and grasping hands with teeny little fingernails – we are hardwired to find our biological family attractive.

I tell the audience that bushbaby babies, adorable as they undoubtedly are, are pretty much superfluous in the great ecological scheme of things – and immediately, I can feel the mood in the lecture room change. *You'll be telling us that pandas are a complete waste of bamboo next!* I merely point out that without bees, flies and beetles, the world would be totally unrecognisable. Flowering plants would go unpollinated, decaying material and dung would not be recycled and a legion of animals, bushbabies included, would not have anything to eat. The whole ecological balance of the Earth is completely dependent on there being an extremely large number of insects, and it has been this way for a long time indeed. No insects? No bushbabies. No you. No me.

The case of the Californian Condor illustrates just how easy it can be to miss what's right under your nose. In the twentieth century, through a combination of poaching, lead poisoning and habitat destruction, the Californian Condor, the biggest bird in North America, faced a very uncertain future. It could easily have gone extinct but, in 1987, when there were only 22 individuals left in the wild, a bold conservation plan was put into action. This majestic bird was worth saving. It was decided that all the wild individuals should be caught to allow a programme of captive breeding to take place. The birds were very well looked after, and this included making sure that they were free of parasites. The birds were deloused using pesticide treatments, and a unique species of bird louse that had only been scientifically described for the first time in 1963 was lost forever. There was no evidence that the lice caused the birds any discomfort, feeding as they do on feather fragments and skin flakes. When vertebrate species become endangered, public interest and research efforts go up, but when insects become very rare or endangered, they are simply forgotten about. It's high time we focused on the things that really matter. Just because they're small, doesn't mean insects are of no importance.

Early days

It's hard to pinpoint exactly when I fell in love with insects. As a child I loved looking at the distinctive black and yellow caterpillars on the nasturtiums in my grandmother's Glasgow garden. I watched them devouring the leaves and eventually pupating to become Large White butterflies. There was always something interesting to find among the undergrowth, and ever since then, my eyes have been drawn to the little lives of tiny creatures. A remarkably prescient primary school teacher once wrote in a school report that a fly going past would distract me from whatever it was I was supposed to be doing. I am glad to say that I still find myself distracted by flies and other insects, and I cannot imagine being interested in anything that is not part of the natural world.

When I was at primary school in the 1960s, the BBC broadcast numerous programmes for schools. There was a series about prehistory and ancient history called *How Things Began*. In one episode there were two children called George and Alice, who discovered what life was like in the seas of the Cambrian Period more than 500 million years ago. There must have been some time machine involved. Another episode was all about life in the swampy forests of the Carboniferous Period, 359–299 million years ago, and how the remains of the trees became the coal that we now use. The programme-makers had hit on the idea of bringing the stories to life by using dramatic presentations, and I seem to remember George and Alice crouching behind a fallen tree, watching giant dragonflies wheeling through the air above their heads. I thought it was absolutely marvellous and, as I collected the coal for the kitchen stove in the years that followed, I would be transported back to the steamy depths of a Carboniferous forest teeming with spectacular creatures. It was much later that I realised the astonishing importance of fossils. We do not need a time machine because fossils are a record of the Earth's history that, in parts, is so finely detailed that it is possible to imagine how whole ecosystems must have looked and functioned millions of years before we appeared on the scene.

How many species?

I was born in the mid-1950s – the beginning of the space age, the jet age and the biomolecular age. I grew up to watch the Moon landing broadcast around the world and see probes sent out into the vastness of space. Since then, computers have shrunk in size and grown in power to the point that they are now ubiquitous and indispensable. Medical science has made enormous progress – organ transplantation is now commonplace, immunotherapy has revolutionised cancer treatment and human cloning may be just around the corner. In my lifetime, humans have invented new ways of observing the world. At one end of the spectrum, giant machines provide proof of the existence of the infinitesimally small subatomic particles that make up matter and, at the other end, space telescopes allow us to see over astronomically large distances to image the boundaries of the cosmos and the afterglow left over from the Big Bang. But despite all these enormous advances, we still cannot answer one fundamental question – how many species are there on Earth?

The first significant attempt to answer the question was carried out by the late Terry Erwin of the Smithsonian Institution. He decided that a good way to work out how many species there might be was to blast rainforest trees with a fog of fast-acting insecticide and see what dropped down. To propel the insecticide up into the treetops, he had to do a bit of rope climbing up into the forest canopy, where he deployed a portable machine known as a thermal fogger. To collect the fallout, large sheets or funnel-shaped collection trays were arranged beneath the trees. Erwin picked one species of rainforest tree in Panama and set about his mass collection effort. What he discovered was that there were 1,200 species of beetles (his specialist interest) living on this one tree species. He reckoned that only about 163 species of beetle were specific to the tree species he had collected from. Erwin then did a crude extrapolation, basing his conclusion on the fact that there were another 50,000 species of tropical rainforest trees and it was likely that beetles made up 40% of the entire insect fauna. He concluded that it was highly likely that number of species on Earth had been grossly underestimated, and his estimate of 30 million species caused quite a stir.

It was not long before entomologists around the world were arming themselves with thermal foggers and mist blowers and killing untold numbers of insects to come up with their own estimates of how many species there might be. Some estimates were as high as 100 million, but gradually these figures were revised downwards and today the majority opinion as to how many species of arthropod there might actually be stands somewhere between 8 and 12 million. However, it was recently claimed that 1 trillion species might currently live on Earth, with only one-thousandth of 1% described. This is due to the realisation that soil, sometimes dubbed 'the poor man's rainforest', contains an incredibly diverse and understudied community of bacteria and other microorganisms. The difficulties in quantifying this diversity are compounded by the fact that the majority of bacteria cannot be cultured, and it's rather hard to say exactly what constitutes a bacterial species anyway. Even sticking with insects, we are still a very long way short of the finish line.

Coined in the 1980s, 'biodiversity' has become quite a buzzword. To a biologist the word means the sum of biological variation from the level of genes and species up to ecosystems, but to most people it simply means species richness – that is to say, the number of species found in any particular area or habitat. But measures of diversity should include both the number of species present and the abundance of each. Imagine two gardens. One is mainly mown lawn and decking with a few pot plants, while the other is wild and weedy. A quick and dirty half-hour survey collects 100 specimens from each garden. In the wild garden 10 specimens of 10 different species are recorded. In the tidy bedecked garden, there are also 10 species (they don't have to be the same as in the other garden) but here, 91 specimens belong to one species and the nine other species are represented by just a single individual. Both gardens have the same number of species but the wild and weedy garden is considered to be more diverse because the probability of the next specimen you collect being something different is much higher here than in the other, tidy garden, where there is a very high probability that the next thing you encounter would be yet another specimen of the common species already collected. You can see why most people like to deal with just the number of species – it's simple and understandable. A few species versus lots of species does mean something after all.

If we are going to appreciate the massive part that insects play in the ecology of our planet, we need to understand where they came from and how they became so successful.

In the beginning

A major obstacle to understanding the history of the Earth is the massive mismatch between events that take place on a human timescale and those that occur on a planetary timescale. The Earth is about one-third the age of the universe – approximately 4.5 billion years old. By chance, a football pitch is 45 metres wide. Therefore, very conveniently, a metre – about one big step – is equivalent to 100 million years and 1 centimetre, the width of the nail on my little finger, represents one million years. The early Earth formed by the slow accretion of immense quantities of dust and gas, and the sideline on one side of the pitch is where we start. As we walk along our timeline, across the pitch to the line on the other side, 4.5 billion years away, there are some important events. The first half a billion years were literally hellish. Much of the planet was still molten at this time and was constantly being knocked about by asteroids. This was also the time that a massive object the size of a small planet cannoned into the juvenile Earth, releasing a vast amount of energy and knocking great chunks of both masses into space. Under the gravitational pull of the Earth, this material coalesced to form the Moon, whose presence has, ever since, stabilised and shaped what happens on Earth. Eventually things quietened down, the Earth's surface began to cool and liquid water formed. Despite the desolate and forbidding conditions that had prevailed for hundreds of millions of years, our planet would not remain lifeless for long.

Of course, the whole reason that life was able to arise in the first place was because the orbit of the Earth is situated in what is known as the habitable zone. It is not so far away from the Sun that it would be too cold, like Mars (-28°C), and neither is it too close to the Sun, where it would be much too hot, like Venus (471°C, more than hot enough to melt lead). This has become known as the 'Goldilocks Effect' because our planet, as the nursery story goes, was in just the right place. This bit of serendipity means that water can exist as a gas, a liquid or a solid depending on where it is found. The oceans that formed on Earth soon after it began to cool may also have been topped up by extraterrestrial sources, such as comets and other ice-rich objects that were flying around the young solar system but had not clumped together to form an actual planet. Some of the chemical building blocks of life, including organic molecules, may have arrived from space carried by comets and asteroids. So far, more than 140 biologically relevant molecules from those bodies have been identified. In addition, intact amino acids, the building blocks of proteins, have been found in the remains of meteorites. In one study that simulated the conditions found in deep space, researchers were able to generate small structures that look very much like cell walls – an important step in containing and isolating all the chemical reactions that must take place inside a cell.

The rise of oxygen

Biologically speaking, nothing much happened for 5 or 6 metres from the start of our football-pitch timeline, but there are now the beginnings of life in the fossil record. Single-celled organisms have appeared, but it is not until we reach halfway across the pitch that the atmosphere becomes rich in oxygen. This took place sometime around 2.5 billion years ago, due to the appearance of the first photosynthetic organisms, the best known of which are the cyanobacteria. These ancient microorganisms lived in shallow seas and had evolved the biochemical trick of making their own food by capturing the Sun's energy. Their generation of oxygen as a waste product was to change the world completely. Cyano-bacteria grew in thin, mat-like layers, binding grains of the substrate together. The upshot of all this oxygen being pumped out was that things got pretty toxic for all the anaerobic microorganisms that already existed and, as levels of oxygen in the atmosphere increased, the ancient bacteria for which oxygen was pure poison were relegated to peculiar refuges. Very few multicellular organisms today are anaer-obes, and those that are find themselves confined to deep-sea habitats, where they use hydrogen as an energy source. As far as we can tell, the trick of capturing the energy from the Sun occurred only once, and the descendants of these early cyanobacteria are still with us today – trees, plants and anything that photosynthesises.

It took quite a while for oxygen to become abundant in the atmosphere because as soon as it was produced, it was immediately removed by reacting chemically with various rocks and minerals. But eventually, after millions of years, there was enough surplus being produced that oxygen began to accumulate. Some of the oxygen became ozone in the upper atmosphere and this meant that, for the first time in the Earth's history, the harmful effects of ultraviolet radiation were greatly reduced.

Many cells are better than one

Life on Earth remained generally small and microscopic until about 600 million years ago – a mere six steps from the sideline on the far side of the pitch – when complex, multicellular life appears. These soft-bodied organisms did not fossilise very well and they were not really like anything that exists today. There were worm-like creatures with segmented bodies, frond-like things and rounded or oval structures that might all have been filter-feeders of some kind.

Complex organisms are made up of different types of cells. Multicellularity, which has evolved many times, has many advantages. For a start, multicellular organisms can get much bigger, and there are economies of scale to be had from using different cells to carry out specific functions. The division of labour itself leads to more specialisation. Multicelled predators would do better and this, in turn, might have brought about a similar response in their prey. But multicellularity relies above all else on communication and cooperation between cells. There would be no point if a nerve cell stopped being a nerve cell or if gut cells decided they might do better ploughing their own furrow somewhere else. And, of course, all cells need to know when they are surplus to requirements – cell death needs to be built into the system. Cancer is an example of what can happen when the strict regulation of cells breaks down and cells, once working for the common good, start behaving like independent agents looking out for themselves. Nearly three years ago a single mutation in a skin cell of my right heel caused it to begin to reproduce uncontrollably – the cells simply didn't know when to die. Fortunately for me the science of genetics was able to pinpoint exactly what that mutation was and a number of drugs became available to block the metabolic pathways within the affected cells. Three cheers for science and for the power of the human imagination and all done at impressive speed given that the structure and function of DNA itself was barely understood in the year I was born.

Some early multicellular life forms were radially symmetrical; that is to say, their bodies showed a repeating pattern around a central axis, like jellyfish and sea anemones. This body plans works pretty well in a marine environment and these creatures spent their time bumbling about looking for food in the sea or fixed to a substrate waiting for food to come their way. But a much better body plan, and one that is seen in the vast majority of organisms today, from earthworms to elephants and bumble bees to birds, is to be bilaterally symmetrical along the longitudinal plane. The first fossil animals with bilateral symmetry would become the predominant life form. Having a clearly defined front and back gives an organism a clear direction of travel. The front part of the body becomes endowed with various sensory organs to measure different aspects of the environment ahead, and the mouth is well placed to consume whatever food may be encountered. Nervous tissue to deal with all this sensory input becomes concentrated at the front end, and gradually the various parts of the body become specialised for different sorts of tasks. Further minor modifications allow for various limbs to provide propulsion and other appendages for food gathering and processing. Now, at last, there is a world-beating pattern that evolution can endlessly tweak and refine.

The creatures from the blue lagoon

Things really start to get interesting about 5.5 metres from the end of our football-pitch timeline (550 million years ago). The Earth's oceans fill up with all manner of complex organisms. The Cambrian explosion, as it is known, saw the relatively sudden appearance in the fossil record of the ancestors of all species alive today. The reason we know a lot more about the last 500 million years or so of Earth's history is mainly because many species developed hard parts reinforced with calcium carbonate, perhaps in response to increased levels of predation. This meant that their remains fossilised much more easily, and this has given us a much more detailed picture of what was going on. This is where the arthropod story really takes off.

Popularly known as the Big Bang of biology, the surge in fauna of Cambrian Period has been well characterised thanks also to the discovery of some exceptional fossil deposits around the world in which the preservation is so fine that minute details of the animals can be seen. These deposits may have formed when a sudden collapse of fine sediment buried the creatures beneath instantaneously. In the right conditions, where oxygen was excluded and any decay stopped, the buried creatures would remain intact long enough for a good impression to be left in the sedimentary rock that it would become part of. The finer the particles of sediment, the finer will be the resulting fossil – rather like the degree of resolution in a digital image. These special fossil finds have given us a series of snapshots of times long ago. With a little imagination, we can swim in the warm Cambrian seas and marvel at the diversity of trilobites and other strange arthropod creatures.

Up to this point every living thing is marine, but 4.75 metres from the end of our football-pitch timeline (475 million years ago), the first land plants appear, accompanied by some of the creatures from the blue lagoons. These creatures became terrestrial and, a short time later (in geological terms), they took to the skies. These first terrestrial arthropods did not emerge suddenly from the sea to live on the land, but probably made a very gradual progression by first living around the fringes of permanently wet coastal areas in habitats such as mats of algae and biofilms growing on rocks. Numerous trackways and trace fossils showing parallel rows of marks made by legs, with a central line of broken marks perhaps created by the end of an abdomen, have been interpreted as having been the steps of walking arthropods as they moved along the shoreline. They might have accompanied the spread of algae and other species as they, too, slowly colonised the land. A common view has been that the invasion of the land by early arthropods took place after the appearance of land plants, but it might well have happened more or less at the same time, as some fossil finds suggest that early food chains comprising scavengers, herbivores and predators were already in place. But what-ever the exact timing, the early arthropods emerged from the sea and made the transition to land, probably in order to escape a marine environment full of other species wanting to eat them. They may have had an amphibious habit of moving between scattered, salty pools. They were without doubt small and fast-breeding creatures, and the insects we know today are their descendants.

Of course, along our timeline there have been extinction events – some of them very large indeed, such as the one that marks the end of the Permian Period, only 2.5 metres (around 250 million years ago) from the far endline, when a great majority of the Earth's organisms perished as a result of massive climatic disruption resulting from volcanism releas-ing vast quantities of carbon dioxide into the atmosphere. Despite this, every living species today had an ancestor that made it through this and other cataclysmic events.

The dinosaurs, the largest land animals ever to have lived on Earth, had dominated their environment for 135 million years, and were seen off by the impact of a large asteroid at least 10 kilometres in diameter smashing into the Earth about 66 million years ago. The impact left a crater 180 kilometres wide and resulted in massive tidal waves, fires and shock waves that triggered widespread earthquakes and volcanism. Acid rain and poor sunlight would have destroyed ecosystems worldwide and caused the extinction of three quarters of the plant and animal species on Earth at the time. Large creatures did not fare well, but small species survived. I like to imagine a small shrew-like creature startled by the massive fireball brighter than a thousand suns, running for the cover in an underground burrow. If they had not, human beings would not exist.

We have now walked almost the whole way across our 45-metre-wide football pitch and the first hominid species that walked upright appears just 3 centimetres (3 million years ago) – about the width of a fat thumb – from the line on the other side. Anatomically, modern humans appear in the last 2 millimetres, the width of a 2-pence coin from the opposite side of the pitch … that's us. We really are the new kids on the block. The whole of recorded human history occupies a paltry 5,000 years, which on our football pitch timeline is about half the thickness of a sheet of gold leaf.

You could imagine that the appearance of life on Earth is a completely unique occurrence – a colossal fluke. Or perhaps life is very common indeed. It has been suggested that as many as 4% of the billions of stars in our own galaxy have rocky planets orbiting them within the habitable zone where water might be available. If you factor in the two trillion or so other galaxies in the universe, you cannot escape the conclusion that that life must exist in innumerable other places.

There are good reasons why insects are so abundant on Earth. They have a versatile, lightweight and waterproof exoskeleton which protects them and keeps them from drying up. Being generally small animals, they can occupy a much larger percentage of any ecological space available. They have an efficient nervous system with an effective blood–brain barrier to keep their vital nervous system insulated from biochemical fluctuations within their fluid-filled bodies and they have absolutely phenomenal powers of reproduction. But for all their jaw-dropping diversity, insects are remarkably similar in overall design – inside and out. The ancient blueprint, half a billion years old or more, has proved a good one and evolution has tweaked it over and over again to produce a multitude of variants based on three main sections: the head, thorax and abdomen. If you were given the task of making a self-sustaining, self-replicating piece of miniature machinery that could survive in widely differing environments, it would necessarily look a lot like an insect.

SIR DAVID ATTENBOROUGH

A wild life well lived

Sir David Attenborough has long held a fascination with the history of life on our planet, and I have been inspired by his work over the years. We spoke about the vital role of insects, and how we can protect the future of life on Earth for animals and humans alike.

Sir David is about as well known a personality as it is possible to be. When I interviewed applicants for a place at Oxford University to study biological sciences, I used to break the ice by asking them what first made them interested in biology. The candidates would regularly tell me about some documentary or other presented by David Attenborough. It would remind me of my first enduring memory of a natural history programme – a short film narrated by Attenborough, not about insects but about a pair of mating garden spiders. I sat in front of our small black and white television set, completely absorbed by the unfolding drama. Having filled his paired inseminating organs, the pedipalps, with sperm, the male spider nervously approached the much larger female. Only by observing the precise ritual behaviours determined by evolution would he be able to avoid being eaten and successfully couple with the female. It really was nail-biting stuff. Nearly 40 years later, I acted as the chief scientific consultant for Sir David's BBC television series *Life in the Undergrowth*, and we have met up on several occasions since. It is often said that you should never meet your heroes in case they turn out to be a disappointment, but I am happy to say that my meetings with Sir David over the years have been delightful and inspiring.

I wanted to know not just if he had had a fascination for animals from an early age but also if insects especially had featured largely. I was a little surprised to learn that insects had not been uppermost in his mind and that the things that fascinated him as a young boy were fossils.

'I grew up in the Midlands, where there's a lovely oolitic limestone full of ammonites and little things called *Rhynchonella*, small little things about the size of a hazelnut and with a zigzag jaw between the two valves. I started collecting them and began to form a collection. I started to work out that some of them were the same and some of them were different; what the differences were, which were common and which weren't and why. It fascinated me – but not in the early years. In the early years, it was seeing how many you could collect and what the variations were. That was what took me out to Eastern Leicestershire on my bicycle, day after day after day during the summer.'

David was lucky to live in the right place and at the right time.

'There were lots of disused quarries and, once exposed, soft, erodible rock like limestone weathers away quite quickly. They stood out like little walnuts on a rock face. You could knock them off with just one tap of the hammer but it isn't like that anymore. Years later, I went back to the quarry and couldn't find anything because it had become overgrown.'

In the end he amassed quite a sizeable collection of fossil material, and I asked him if he still had it in his possession.

'No, unfortunately not. My father was head of a university college in Leicester, which opened a geology department during my teens, and I gave them the whole thing! They used it to teach students or, perhaps they threw it away!'

I sincerely hope they held on to them.

David won a scholarship to Cambridge, where he wanted to study zoology and botany, but a certain amount of parental pressure sought to persuade him that there was no future in either subject, and that geology might be the thing to study. After all, he had a lot of paleontological knowledge already and if he was to have a good career, the oil industry offered excellent prospects. David explained that being able to date rocks precisely using the microfossils they contained was all very well, but he would also have had to study things like mineralogy and X-ray crystallography, which he admits did not interest him at all.

After graduating from zoology and botany in 1947, David served in the Royal Navy and then wondered what he would do next. He didn't feel like going back into academia and instead wanted to earn a living, perhaps using his scientific qualifications. He admitted, in typical self-deprecatory fashion, that his qualifications were a minimum.

He got a job as a science editor in educational publishing but did not enjoy it one bit. 'I spent my days putting in a comma and, on good days, taking it out' he recalls. One day he replied to an advertisement in *The Times* for the position of Radio Producer. He didn't get the job, but a couple of weeks later, he received another letter asking him if he had heard about a new thing the BBC were starting in North London. It might not amount to much, the letter went on, but would he be interested in finding out a bit more about it? That 'thing' was television. 'I thought, well, why not? What I'm doing at the moment is utterly boring. I got taken on and I've been there ever since.'

I wanted to talk to him about *Life in the Undergrowth*, his five-part series on insects and other invertebrates. He had already made a number of highly successful and award-winning documentaries, including *Life on Earth*, *The Trials of Life* and *The Private Life of Plants*. Had he been concerned that it might not attract many viewers? After all, people don't generally find insects and spiders that interesting. He agreed that birds and mammals were indeed more popular among viewers, but said that it would be impossible to simply ignore the terrestrial invertebrates, as they are the most important animal group. His problem was what to call the series. He couldn't call it *The Life of Insects*, as many other things, such as spiders and earthworms, would be featured as well. Then, one night, he had one of those 'eureka' moments when the answer just pops in your head: *Life in the Undergrowth*! As he spoke, I imagined him leaping from the bath, shouting 'I've got it!'

They worked away on the series for a couple of years and were rather surprised to receive a memo from a committee whose job it was to advise on programme titles. They had done some market research and had decided that the word 'undergrowth' might have unpleasant connotations – like 'underwear'. Would David and the senior team like to come for a meeting and kick some ideas around? No, they would not, thank you.

Much later, when the programme was finished and just about to be advertised in the Radio Times, another note arrived from the committee of programme titles. The committee had deliberated long and hard and thought that *Life of the Multi-legs* would be a good title!

Happily, they stuck to *Life in the Undergrowth*, and I thought it was absolutely marvellous. It was one of the first programmes that really brought the life of invertebrate animals to a wider audience. Their importance in the functioning of global ecosystems is unparalleled and yet here we are, 20 years on, still trying to get the same message across. I can still remember hearing the lines Sir David spoke at the end of the final episode:

> *If we and the rest of the backboned animals were to disappear overnight, the rest of the world would get on pretty well. But if the invertebrates were to disappear, the land's ecosystems would collapse. The soil would*

lose its fertility. Many of the plants would no longer be pollinated. Lots of animals, amphibians, reptiles, birds, mammals would have nothing to eat. And our fields and pastures would be covered with dung and carrion. These small creatures are within a few inches of our feet, wherever we go on land – but often, they're disregarded. We would do very well to remember them.

I wondered if he thought that we had lost our connection with the natural world. Sir David agreed and worried that because we have become so urbanised, we have become cut off from the natural world. He did recognise his own limitations, such as his ignorance of birdsong and that fact that he probably knows as much about the natural history of Borneo than he does about the English countryside.

However, he is convinced, as am I, of the importance of engaging children as early as possible. He told me about the time he took his five-year-old godson to a field to see what they could find. They looked under some stones and the boy exclaimed 'Oh, look, what a treasure. A slug!' David went on to say, 'And, of course, he's right.' The boy asked so many questions: 'Can it see? What are those things sticking out at the front? How does it eat? How does it move?'

Perhaps David remembered his own fascination with such discoveries. He believes that when you're four or five years old, you have no prejudice, and that innocence of eye is so precious.

Had he ever been in fear for his life on some big jungle expedition? David admitted that the only time he felt in any kind of danger was when he or the filming team failed to recognise signals from the animal they were approaching. 'It's part of an inheritance from our early Palaeolithic past, that we should recognise whether an animal is angry or not,' he explained. I reminded him of the famous sequence of him playing with gorillas – one serious swipe from that silverback gorilla could have ended his career right there. David was pretty sanguine about the encounter and said that danger signals are easy to see and, as long as you behave yourself and don't take liberties, there won't be a problem. If you get it wrong, it could be disastrous.

I like to think the big male gorilla simply knew who he was sitting beside.

Sir David has been in the business of natural history for 70 years and realises that the world is changing. He gets 50–70 letters every day, many from people in their teens and early twenties. He told me that some are simply addressed to 'Attenborough, London'. How cool is that?

'Mass communication has changed everything, and now people all over the world understand the possibilities of extinction and the dreadful things that human beings have done to the world.' They are aware of the dangers and are desperate to do something about it. He's pretty clear on what we should do:

'Don't waste food, fuel, gas, electricity, space, plastic or paper. If you live in a more economical way, you are helping to restore the balance. The huge problem, of course, is the increasing population so we must look at the world from an international point of view. We have to get together and say, "Okay, we have to work out a plan that every nation can subscribe to, and every nation will benefit from so we can actually stop these threats to our planet." So while you must not waste things, you also have to support the politicians who are going to attend these conferences and fight for an international agreement.'

I asked David about how, in the early days of television, there was a tendency to not show things as they really were – nature without the nasty bits. I mentioned the state of the oceans. He said that these international commons, which belong to no one, are overfished to a dreadful degree, but he is optimistic that solutions can be found: 'We need to stop arguing and come to some kind of agreement.' He sounded positive and reminded me of the time we agreed to stop killing whales. He admitted there were still loopholes but, as a result of that agreement, there are now more whales in the oceans. He added: 'When I first tried to film blue whales 50 or even 30 years ago, it was impossible.'

The blue whale's sole food source is krill, a small shrimp-like creature. Although the estimated biomass of krill is around 400 million tonnes, it is thought that over half of that amount is eaten by marine animals. We currently only take less than half a million tonnes for use as fish and pet food but, if we started to harvest krill on an industrial scale as a human food source, we will endanger whales once again, as well as many other species.

What about the laudable aim for a third of the Earth to remain as wilderness? There will be arguments about whose wilderness we are going to keep and what they will be paid in return. We agreed that it was going to be very complicated. We talked about the wonders of the tropical rainforests and how many species are yet to be discovered. He said that we would never be able to catalogue all the species of life on Earth, but maybe we don't have to. Maybe we should just save as much of the habitat as possible so that it remains intact and safe.

'If you knock down the Amazon rainforest, you will completely disrupt the climatic patterns of the world and the consequences will be felt all over the globe. What is needed is an international agreement that is recognised.' David said he thought the rudiments of such agreements are already in place. 'And some of us have to pay. I mean, developed nations of the world have done very well, exploiting stuff from the less developed nations of the world. And that can't go on.'

The greatest accolade a biologist can receive is to have a species named in their honour – I have five or six species of insect that bear my name, but David has had more than 40 species of animals and plants named in his honour. He reminded me, with evident glee, of a genus of the extinct pliosaur, *Attenborosaurus*, which swam in the tropical seas that are now the south coast of Britain.

David recognised the vital importance of taxonomic expertise. He said, 'What's required when someone comes along and says, "Is this beetle with these scarlet flashes on it or a new species?"? Where would you find the man or woman who's going to tell you the answer to that question? In a zoology lab somewhere, but they're much rarer than the animal itself!'

Sir David Attenborough's words carry a lot of weight and people listen. What did he think of his life and what he has achieved?

'I can't imagine how anyone could be as lucky as I have. I'm completely out of date in terms of science. I'm, in a sense, a butterfly flitting from flower to flower, very superficial – and I'm well aware of that. If it served a function to be superficial and have that breadth of experience, then it's absolutely great. As a 12-year-old, I could not possibly imagine that I would lead the life I do, to be able to go anywhere I wanted to in the world. I wouldn't mind going to the Gobi Desert, but there are so very few animals there. I feel that drawing attention to what's happening to the variety of the natural world is an important thing to do. If I've contributed to that, then I've slightly justified having this wonderful opportunity. I'm beyond grateful for it.'

Sir David has indeed been very lucky, but then so have the many millions of people who have watched his programmes and been inspired by the wonders of the natural world.

• • •

In the next chapter we will examine the details of the insect blueprint to discover what it is about these small creatures that made them the most phenomenally successful, enduring and diverse group of organisms our planet has ever seen.

Chapter 2

BRILLIANT BODIES

Designed by nature

The remarkable diversity of insects and their spectacular success is only possible because of their incomparable design. I wish there was a word that did not suggest a designer or creator other than the persistent and pervasive power of evolution through natural selection. Millions of years of refinement has produced the ultimate survivor, and it would be difficult to imagine a more adaptable and versatile animal than an insect.

Many of the ancestors of insects had more segments and more legs, but no single region of the body was particularly specialised except the head. The trouble with this arrangement is that it is not very efficient, and is partly why there are a hundred times more insect species alive today than there are centipedes and millipedes. Over time a more advantageous arrangement would evolve, where segments of the body became grouped into different functional sections. The advantage of this amalgamation of segments is that each group of segments could become a specialised region for different tasks. Evolution has tweaked the ancient insect blueprint over hundreds of millions of years to produce a multitude of variants based on three main body sections: the head, the thorax and the abdomen.

The insect head is the command centre. It is formed from six fused segments and carries a pair of compound eyes comprising a large number of individual light-gathering units, secondary light-receptive organs called ocelli and a pair of antennae. It also carries the mouth-parts, which are variously modified to allow the sucking of liquids or the chewing of solid food. The thorax, made up of three segments, is the powerhouse of the insect. Each segment bears a pair of legs and the rear two segments typically each have a pair of wings. The abdomen, which is usually made up of 11 segments, contains the diges-tive system and reproductive organs. And that's it! Head. Thorax with legs and wings. Abdomen. It's a compact and easily modifiable arrangement.

In life, the internal organs of insects are immersed in a plasma-like fluid called haemolymph that acts as a multipur-pose chemical-transport system. This arthropod version of blood is pumped from the back of the body to the front by an open-ended, tube-like heart lying along the dorsal surface.

The digestive system of insects is pretty much like that of many other animals – essentially a tube. The front part of the gut has special regions for grinding and storing food; the midgut is the main area for enzyme production and the absorption of nutrients; and the hindgut gets rids of waste.

Like all animals, insects need carbohydrates, fats, proteins and vitamins to survive and grow. Plant foods, especially sap, can be very low in protein, and sap-sucking insects may have to eat large amounts to meet their requirements. Herbivorous insects can be quite fussy, and most are restricted to eating one plant or a few closely related ones. This is not true of the desert locust. One of the most damaging insects on the planet, this species is capable of forming immense, crop-devouring swarms comprising billions of individuals. They have a reputation of being able to eat just about anything, but the reality is that they are rather careful what they eat – it turns out that they are not gluttons, but gourmets. Research on their feeding behaviour shows that they regulate their food intake very well (much better than we do, if left to our own devices) and eat fats, carbohydrates and especially protein in accordance with what they need at any time.

No one knows more about insect nutrition than entomologist Professor Steve Simpson, and I spoke to him about his groundbreaking work on locust feeding habits.

PROFESSOR STEVE SIMPSON

Life lessons from locusts

I have known Steve Simpson for more than 35 years, both as a colleague at Oxford University Museum of Natural History and as a friend. I joined the staff of the Hope Entomological Collections in 1984, and Steve arrived two years later to take up the position of Curator. I remember those years as among the happiest of my life.

Steve has made it his life's work studying the physiology and behaviour of insects. Using locusts as a model organism, he and his collaborators have uncovered some basic nutritional rules that apply to all animals. Steve is a consummate researcher, teacher and team builder and, during his time at Oxford, I felt very privileged indeed to be his wingman.

He now heads up the multidisciplinary Charles Perkins Centre in Sydney, Australia.

If anyone can explain the importance and relevance of studying insects, it's Steve but before I got into the scientific substance of the interview, I wanted to know more about his childhood growing up in Australia.

'I think I was born an entomologist. My godmother asked me as a three-year-old: "What do you want to be when you grow up?" And I said to her, "I want to be an entomologist." I've been uttering the word "entomology" since I was a small child!

'In those very early days, my mum (who had grown up in the Bush just outside of Sydney) had a love for entomology and the natural world, and encouraged my English father to get into it too. He had virtually never seen an insect, or at least anything larger than a cockroach, in the midwest of England. She used to take him out during the late spring/ early summer evenings to collect greengrocer cicadas as they were emerging and climbing the tree trunks in our front garden. They would carefully collect them and hang them on my curtains, so that when I woke up in the morning, I'd have these beautiful green, large cicadas hanging there. They were, and still are, my favourite animal.'

Steve mentioned that there are fewer cicadas around today than when he was growing up – I wanted to find out how the makeup of insects has changed over the past few decades in Australia.

'The cicadas' story is a little nuanced. Their numbers go up and down seasonally – some years there's larger emergences than others, and that's because they're synchronised in their development below ground.

'When I was a kid, there were different values placed on different species among the kids in the neighbourhood. Different colour morphs or species of cicada got you more bargaining power for whatever token you were using to negotiate with the other kids. There was a black species, which was incredibly rare. If you found one, it was worth 10 greengrocer cicadas. Nowadays, you'd find far more of the black ones than the green ones.

'There's been a change in the relative frequency of different species in urban environments, but that's accompanied by several other changes. Urban Australian habitats are now much richer in birds than they used to be, partly because they've moved out of the surrounding countryside into urban environments. Insect numbers have declined overall, but some seem to have become more abundant.'

Steve spoke about how insects featured prominently in his childhood and how impressed he was at seeing the Giant Valanga grasshopper (Valanga irregularis) *for the first time. This spectacular insect is the largest grasshopper in Australia.*

'I grew up in Melbourne, but we moved to the subtropics in Brisbane when I was nine. We moved into a house with no screens on the windows and a light overhanging the dinner table. All manner of extraordinary things used to land on our dinner. My dad was pretty alarmed by the size of some of these!

'I had a real love for grasshoppers, and I still do. The ones I would collect were all pretty small. Then we moved to Queensland, and I remember walking into the back garden one day and seeing a Giant Valanga – a species of a very large grasshopper related to the desert locust in Africa, one that I've spent quite a lot of my career looking at. It was like seeing a dinosaur. I just couldn't believe it. I crept up on it and tried to catch it but it leapt off just as my quivering hand was approaching, and flew into the neighbour's garden. It was a terrible, terrible loss so I set out to try and find more of them. When they're nymphs, they are brightly coloured like the desert locust, and they feed on vegetation growing in people's tropical gardens. My brother and I used to go out and collect them. We spent hours walking around in the boiling sun, to the point that one day he ended up with sunstroke.'

I asked whether he thought Australians have a more relaxed attitude to insects than Brits. Most people are comfortable around hover flies and bees, but if we had giant grasshoppers here in the UK, I imagine there'd be a run on pesticide spray guns.

'Australians have a much more relaxed attitude towards their garden's wildlife than the Brits. All of the extraordinary insects were just part of the yard – the local fauna – and part of the interest.

   'The other thing that we had in large numbers in our yards in Brisbane were cane toads. These invasive, hideous amphibians from Central America were originally introduced to North Queensland to control the Grey-backed Cane Beetle. They introduced this cane toad thinking it would act as a biological control agent. Having had remarkable success with the *Cactoblastis* moth, which cleared the *Opuntia* (prickly pear) that had overrun much of the Australian environment – an incredible example of entomological biocontrol – they thought, "this is easy" and introduced the cane toad. Of course, it turned into one of the great invasive pests of the planet. At night when you went out, you could tread on a cane toad with your bare feet.'

Some people might imagine that a childhood passion for insects is not going to translate into a proper job for a grown-up, but they'd be wrong – Steve made a significant research career out of it.

'You might not have predicted this but studying insects has led to a fundamental understanding of the causes of the human obesity epidemic. That sounds kind of weird, but it's true. I had a real interest in how animals, particularly insects, know what to eat and when to eat it. My honours project looked at newly emerged sheep blow fly maggots, *Lucilia cuprina*. Rather than being a fly, it has evolved one step further and infests in living sheep. They lay their eggs deep in the foetid crotch wool of the sheep. A whole bunch of little maggots emerge at once, and they scrape away at the skin of the sheep until they cause a lesion and, ultimately, eat the sheep alive.'

This horrendous condition is known as fly strike, and in Australia it became such a severe problem that it threatened the whole sheep-rearing industry. Steve's undergraduate project looked at the feeding behaviour of the maggot and examined the details of their feeding choices. Later, he came to London to take up a PhD position looking at locust feeding biology.

'In the UK I studied the same piece of biology, but now with a different animal: locusts. This is a group of acridid grass-hoppers that all share a remarkable capability. If crowded, they quite literally change their behaviour, shape and colour, such that they become swarming rather than solitary grass-hoppers. That switch between the solitary and swarming form is at the heart of the crop-destroying locust problem, which affects about 1 in 10 people on Earth.

'You might think if you are going to look at an animal and try to understand its appetite system, that a locust would be too voracious and non-considered in its nutritional choices to study. And that's true if you take 100 billion locusts, then they'll eat the same amount of food as the population of London within a week. But every single locust is exquisitely able to balance its nutrition. It knows what to eat and when to eat it. My PhD project looked into this problem, and took me all the way from delivering Vaseline enemas to locusts to try and find out whether they had stretch receptors in their recta all the way through to experiments where I was teasing apart some of the behavioural cues that they use and their rhythms of feeding.

'That study then led me to examine a more complex animal to look at its feeding neurophysiology. I started to work with monkeys in an experimental psychology department at Oxford. However, I soon discovered that to get fundamental answers about how an organism made nutritional decisions, I needed to go back to my locust system in zoology, which is when I really started to uncover the fundamentals of appetite.

'It turns out that locusts and other insects don't just feel hungry or full. They have specific appetites for different nutrients. That early discovery, which then led to trying to understand the mechanisms that control those sorts of separate appetites, turned out to be crucial. Myself and another PhD student, David Raubenheimer, developed a new way of looking at nutrition, based on our study of locust feeding behaviour. That led us to discover that not only do locusts and other insects have specific appetites for different nutrients, so do all animals – including humans.

'We discovered that if you put those appetites together, in an environment where they're forced to compete with one another, protein is the dominant appetite. We found that the appetite for fats, carbohydrates and protein don't work *together* to help the organism balance its diet as they do in natural food environments, but rather, in an imbalanced nutritional environment – such as the modern human food environment – then these appetites are forced to compete – with protein dominating.'

This all makes sense because if you're going to grow and reproduce, you need to eat protein.

'You need protein because it contains nitrogen, an essential element when it comes to building and maintaining new tissues and reproducing – and everything else that matters in life! Calories and nitrogen (plus all the other micronutrients) are embedded in the appetite systems possessed by all animals, including ourselves.

'We then went on to propose that perhaps a dilution of protein in the human food supply – which is what's happened over the last 50 years or so as a result of the incorporation of vast quantities of highly processed fats and carbohydrates in the form of ultra-processed foods – is what's driving overconsumption of total energy, which leads to obesity and everything else that goes with it.'

Steve is basically arguing that as human beings, we are faced with a modern smorgasbord of processed food which has attractive and delicious elements, but it hasn't got much protein. Ours brains are telling us that we're not getting enough protein, so we compensate by eating more (protein-deficient food) to reach our daily target. It gets worse. Steve explained that highly processed foods are sometimes made to taste like protein, while actually only containing fat and carbohydrates.

'A barbecue-flavoured potato chip has all the sensory cues of a high-protein food and all the flavours that we've learnt to associate with protein. If you're seeking protein, you'll eat them but get no protein. Your protein appetite will be saying, "You haven't eaten any, keep going!" As a result, you're going to eat more calories to reach your target.

'This idea is called the "protein leverage hypothesis" and was published in 2005. It has proven to be consistent with the evidence across global populations for the emergence and the continuation of the human obesity epidemic. It has had profound implications for human health – and came from watching locusts make food choices.'

Steve has shown that you don't have to focus on large, hairy animals if you want answers and you want them quickly, and that using insects as model organisms is a very good idea indeed.

We talked about the food industry and how it seems to manipulate consumers. I said I feel rather annoyed that we can be so easily conned. Well, perhaps 'conned' was the wrong word. The food industry may not have known what Steve Simpson and his colleagues found out, but they certainly knew what sold well and were happy to produce more and more of the same. They have been making money at the cost of our collective health.

'These foods have been tailored to be irresistible. If you take the combination of fat, sugar and salt that you find in many of these highly processed foods, you've got a supernormal stimulus in the same way that, as a very small boy coming across a giant grasshopper in the back garden in Brisbane, I was totally elated. The same thing happens with food.

'Very seldom in the natural world have foods evolved for the purpose of being eaten. Generally, it's the other way round: your foods try *not* to be eaten. They run away very quickly or become poisonous or spiny. Very few things have evolved specifically to be eaten. Flowers *have* evolved for the sole purpose of being eaten by pollinating insects – a fantastically productive relationship for everybody concerned. Whereas ultra-processed foods have been designed to be irresistible – to be eaten for very different reasons – the hunger for profit rather than the hunger for nutrients.'

Around a decade ago, Steve left Oxford to go to Sydney, where he is now the academic director of the multidisciplinary Charles Perkins Centre. He told me how this came about.

'Ten years ago, I set up the Charles Perkins Centre at the University of Sydney. It's a multidisciplinary initiative with a mission to try to understand and fix the global pandemic of obesity and metabolic disease. And the answer stems back to locusts.

'At the Centre, we want to encourage people to come together across many different disciplines. That's a hard thing to do because people speak different "languages" when it comes to scientific or academic disciplines, and we needed everybody to come together – from the arts, humanities, social sciences, physical sciences, the life and medical sciences – to solve this huge, complicated problem. So we founded an institutional model on the same principles we discovered when trying to understand why it is that locusts swarm. These models take advantage of the idea that when lots of entities interact with one another – whether that be researchers or locusts in a swarm – extraordinary things emerge out of those interactions, if structured correctly. And that set of principles I first discovered when working with locusts.

'My move into trying to understand swarming in locusts was stimulated by the fact that dieldrin – the principal chemical used to control locust swarms – was banned across the globe in the 1980s because it was causing terrible ecological damage.

'After that, the United Nations released some funding to try and find better ways of predicting, managing and controlling outbreaks of the desert locust, the one of biblical fame. We soon discovered not only the mechanisms by which they shift from one phase to the other when they're crowded, but also the implications that has for the emergence of mass migration behaviour, which involved demonstrating that individual locusts in a crowd have some really simple local interaction rules.

'Locusts actually try to avoid one another but, if they get too close, they align with others within a certain distance and move together. That simple behaviour can lead to a sudden mass collective movement – as if of a single mind, millions of locusts will start marching. So when the winged adults fly together in a swarm with no leader, it's simply an emergent property of local interactions.

'Similarly, if you put people together in groupings that you wouldn't find within a typical disciplinary environment, like a department or a faculty, it will lead to extraordinary things. You can't predict them, but they'll be novel and you'll be able to discover new things that you wouldn't have discovered using a more conventional way of doing things. If entomologists never interact with behaviourists or physiologists or any other sort of person, they will be constrained. You shouldn't have departments of specific interest because they're all going to be very insular, very inward-looking.'

Steve's intergrated approach encourages people to cooperate rather than plough their own furrow – and this was discovered through watching the behaviour of insects as a system. I asked Steve whether he was aware of any other group of animal from which we have learnt more about genetics, physiology and behaviour.

'Some of the foundational principles in the understanding of behaviour obviously emerged out of insects, whether that be Karl von Frisch with his work on bees or Niko Tinbergen with his work on wasps' spatial location and how animals navigate and understand where they are in space. These are fundamental principles in genetics, in physiology. Even now, molecular biology and the understanding of human disease processes rely heavily on some of those breakthrough discoveries that were first made.

'These crucial discoveries are helped by the fact that, at a very fundamental level, humans are just flies. There are so many things that humans have in common with the molecular biology of a fruit fly, which means if you can start to explore the processes in an organism as simple as a fly, there's a real chance of translating some into our understanding of humans. I've got a remarkable colleague at the Centre, who takes large genetic screens in human populations, seeks candidate genes that might be associated with a particular disease, then goes back to the fly. Then, with modern molecular genetic techniques, plays with those same genes or their homologues [those that are similar] present in the fly. My colleague then moves back into clinical translation in a mammalian system, looking at the trial of new interventions, which may be genetic, drug or lifestyle interventions. So there is an interplay or dialogue between the insect system, the human system, and clinical outcomes. Insects are centre stage of new discoveries, as well as having given the inspiration that provides the foundation for much of modern biology.'

Six superpowers

To understand why insects are such successful animals, we need to examine a number of key features. Insects have six superpowers that have helped them thrive. Because they are arthropods, they have a protective exoskeleton. They are generally small creatures that survive better than large species. They have an efficient nervous system that suited their transition to dry land, coupled with a fantastic ability to sense the environment. They evolved the power of flight, making them the ultimate colonisers and escape artists. Finally, they are prolific and reproduce at an extraordinary rate.

Advanced armour

The shapes of some insects can seem completely fantastical – so bizarre, in fact, that it is sometimes difficult to imagine how or why they look the way they do. I have been lucky enough to see some of the weirdest-looking of all insects first-hand.

It had a been a long and tiring day involving three flights – London to Chicago, then on to Miami and eventually to the International Airport in Belize. After a four-hour drive I arrived at the Las Cuevas field station, situated in the heart of the Chiquibul Forest Reserve, the largest protected area in Belize. It was August 2001 and I had come to Belize to search for sap-sucking bugs called treehoppers. Related to cicadas and planthoppers, what makes these insects so remarkable is that they are truly masters of disguise.

I had first seen treehoppers a few years before in the entomological collections of the Oxford University Museum of Natural History, where I worked. The polished mahogany cabinet that housed them contained half a dozen drawers of treehoppers. I opened the first drawer and peered at the specimens pinned inside. Treehoppers are quite small insects and I decided to carry the drawers down to my office where I could look at them under a stereo-binocular microscope. I spent the rest of the afternoon happily taking specimen after specimen out of the drawers to examine them. It was clear that they had not been handled in a very long time because some of the pins were hard to pull from the cork base of the drawer. Much of the material seemed to have been collected in the nineteenth and twentieth centuries, with a few additions to the collection in the last 50 years.

Some of these treehoppers looked like plant thorns, while others looked like seeds, but the most interesting specimens in the drawers looked a bit like ants. Treehoppers, of which there are around 3,500 species, are especially diverse in South America, and exhibit some of the most freakish shapes seen in the insect world. In some insects several parts of the body can look leaf-like or stick-like, but with treehoppers, the weird and wonderful modifications that camouflage them are confined to just one part of the body – the back, or dorsal, surface of the first segment of the thorax, which can be greatly enlarged and contorted. Some species have a long projection from which other projections branch off horizontally, each with a small ball attached. In others, the whole thing looks like an ant with its jaws held wide open. These elaborate structures, which often look much bigger than the treehoppers themselves, have no obvious function but to conceal the rest of the treehopper. They look bulky, but they are hollow and lightweight. Whatever the cost to the individual treehopper in terms of the energy required to make such a thing and to carry it around, the advantages must be considerable. Treehoppers may look pretty obvious pinned in neat rows in a museum drawer, but in their shaded rainforest habitat, they can be hard to spot. Their portable camouflage protects them from hungry birds and may even have become part of the way different species recognise a potential mate.

As I gazed down the microscope at the weird shapes, I wondered what might have been the evolutionary path that led from a plain unadorned treehopper to these carnivalesque creatures. Imagine a population of ancestral treehoppers. They are small and unremarkable. By chance, a mutation arises that gives one of the treehoppers a small spine-like process on the back of the first thoracic segment. Because of how it looks, naive birds and other diurnal predators might avoid it more often than not. This individual survives to produce slightly more offspring than the normal-looking treehoppers. Predation is a powerful selective pressure and, over hundreds of generations, the small spine becomes enlarged; in a thousand generations, it can assume all manner of shapes – just so long as each step along the way looks increasingly inedible. These brilliant bodies demonstrate one of the keys to survival: not looking like a predator's next meal.

Another defining feature of arthropods is the covering that separates their insides from the outside. It's called a cuticle and it's made of something called chitin. The cuticle covers the whole of the outside of the insect and also extends inside, where it lines the foregut, the hindgut and even the major branches of the tracheal system, by which means air enters the body. The chitin molecules are embedded in a protein matrix to make a tough composite material. In parts where extra toughness and protection are important, the outer layer of the cuticle can be tanned or hardened. This is an irreversible process caused by adjacent protein chains becoming chemically cross-linked. The resulting material can form the hardest wing cases of beetles as well as claws and mandibles capable of cutting through hardwood and even some metals. Areas that need to remain soft and flexible, such as the joints between leg and body segments, remain soft and pliable. One of the major problems of life on land is that, exposed to sun and wind, you can start to dry out. To combat this, the very outermost layer of an insect's cuticle is waterproofed with a thin layer of wax.

But the cuticle of an insect is markedly different from a simple suit of armour. It is capable of a degree of self-repair and, importantly, forms an exoskeleton, providing anchor points to which the internal musculature is attached. The cuticle also provides protection from the biological warfare mounted against insects by viruses, bacteria and fungi and, in many cases, is able to mount an immune response to these pathogens by producing antibacterial proteins. Modern science has come up with all kinds of composite materials that are both strong and lightweight, but nothing we have manufactured thus far quite matches the versatility and effectiveness of an insect's cuticle.

Of course, having a tough exoskeleton brings its own problems. Vertebrate animals have an internal skeleton which grows as they do. A big drawback for creatures with a skeleton on the outside is that it needs to be renewed from time to time to allow for growth. Some insects moult as few as three or four times during their life, while others may do so 50 or more times. There are dangers in moulting as, for a short time, the protective function of the cuticle is compromised.

If you have ever spent time sitting near water, you may have seen a fully grown dragonfly nymph crawling slowly up a plant stem to begin shedding its skin. The first time you witness an insect moulting to the adult stage, it can seem utterly mysterious and otherworldly. I have never met anyone who has not been fascinated and absorbed by this remarkable process – one that has been repeated trillions upon trillions of times since arthropod animals first arose. It truly is one of the wonders of the natural world.

When moulting begins, the cells of the epidermis draw back from the inner surface of the old cuticle. The surface of the epidermis increases by the cells dividing and, having a greater area, now appears wrinkled. This is important, as the surface area of the epidermis determines the surface area of the cuticle it will eventually secrete. Then, a special fluid – a mixture of enzymes – is secreted into the space between the epidermis and the old cuticle, which begins to be digested and reabsorbed. The fluid cannot attack the toughest parts of the cuticle, which will be eventually shed. Once the epidermal cells have secreted a new cuticle, the process of shedding the old one takes place and the insect must expand the new cuticle before it hardens.

Size matters

Why are insects generally small creatures? Why are there no cat-sized cockroaches or dog-sized dung beetles? You may be very glad there are neither. Although some of the largest insects can feed on some of the smallest vertebrates, there are good mechanical and biological reasons why insects cannot be larger than they are.

The size range of all organisms covers eight orders of magnitude. The smallest bacterial cell, at around 0.3 millionths of a millimetre, is 10 times smaller than the finest strand of spider silk. The largest animal ever to have lived is the blue whale, at about 30 metres in length. In the middle of this range, with an average length of around 3 millimetres, are the insects. There are, of course, insects that are larger and smaller than the 3-millimetre average. Goliath beetles, found in African forests, can be up to 110 millimetres in length. They sit neatly in the human hand and weigh between 60 and 100 grams, which makes them heavier than many birds. A few butterflies and moths have very large wing spans (up to 280 millimetres across) but their bodies are small. The longest species of insect in the world, a stick insect found in Southeast Asia, has a body length of around 320 millimetres – about the length of a human forearm. The smallest insects are species of parasitic wasp that lay their eggs inside the eggs of other insects. Although winged, these minuscule creatures are a little more than 0.2 millimetres from head to tail. This is about half the size of some single-celled organisms, such as an amoeba. One of the smallest beetles in the world today is a tiny feather-winged species that could balance quite easily on the claw tip of a goliath beetle. That insects can use the same body plan across a large range of sizes gives them an unrivalled versatility.

In my travels I have seen some impressively large insects, some of which have tried to take a chunk out of my fingers, but on an East African field trip some years ago, I collected a truly tiny wasp that I thought might take the all-comers record for the smallest insect on Earth. I could not measure it accurately in the field and made the rookie error of mentioning my discovery to a journalist before I had checked my facts. On my return to Oxford, I got phone calls from several local and national newspapers asking to see the wasp. I had to confess that it would take me some time to locate the specimen among the masses of specimen tubes I had brought back. A national paper ran a story the next day with a cartoon of a balding, bespectacled white-coated scientist with a puzzled expression, holding up an empty flask. The caption read: 'I've found the world's smallest insect; Oh dear, where is it?'

That an insect less than a millimetre long can exist and function is a fantastic example of the success of the insect blueprint. The tiny wasp I had collected has exactly the same organs, working in exactly the same way, as the giant spider-hunting wasps I would sometimes see patrolling the savannah to find their next victim. Though the mass of the giant wasp is 1,000 times greater than that of the tiny wasp, they have the same nuts and bolts and are both part of the ecological machinery that makes the world tick.

Insects were not always as small as they are now. There is ample fossil evidence that in the late Carboniferous Period, many arthropods were much bigger than they are today. They might not have been as big as the monsters dreamed up by the makers of horror films, but to our eyes, they would have seemed absolutely huge. There were giant, flat-bodied millipedes, some of which had a body length greater than 2 metres. There were silverfish 6 centimetres long, big enough that if they were running around your bathroom, they would give you quite a turn. Enormous primitive dragonflies with wingspans of 75 centimetres were wheeling through the air, and large cockroach-like creatures scuttled through the humid undergrowth. So why are no very large insects around now?

The amount of oxygen in the atmosphere provides one explanation. The atmosphere today comprises about 21% oxygen, but in the Carboniferous Period, partly as a result of the extensive forests growing at that time, oxygen made up as much as 35% of the atmosphere. Insects have a system of gaseous exchange based on simple diffusion. Along the length of the body there are holes, called spiracles, which are the openings to a labyrinthine system of air-filled tubes called tracheae that get smaller and smaller as they spread throughout the body. The next time you see a caterpillar, look closely and you should see a dark spiracle on the side of each segment. The finest branches of the tracheal system are in intimate contact with individual cells, and it is here that carbon dioxide diffuses out of the cells and oxygen diffuses in. The bigger an insect is, the more difficult it is for oxygen to diffuse all the way to the cells in the interior, but if there is more oxygen present in the atmosphere, insects can get bigger.

Another explanation is that large size may have evolved as a defensive strategy. In the Carboniferous Period, insects pretty much had the place all to themselves; one way to avoid getting eaten was to be bigger than the thing that was about to eat you. This, in turn, would lead to predators getting bigger and so on, resulting in an evolutionary arms race of size. The process would have continued up to a certain point when other factors started to become a problem. Towards the end of the Carboniferous, the appearance of many vertebrate predators such as reptiles, which could get a lot bigger than insects ever could, meant that being a large insect was no longer such a good idea. In any case, the climate was changing and many of the lush forests that existed disappeared to be replaced with much drier landscapes.

Humans are quite large animals and we know, intuitively and from bitter experience, that there are certain things we cannot do. We cannot fly unaided and we cannot jump off high places without the risk of serious injury or death. To really understand the world of insects, you need to imagine how the physical environment affects them. If you live on Earth, you cannot escape the force of gravity, which causes falling objects to accelerate at a rate of 9.8 metres per second squared. Gravity is a constant, but its effects vary according to how big a falling object is. Regardless of differences in mass, all objects accelerate at the same rate unless air resistance affects one more than another. Because of drag force on the objects, they do not accelerate beyond their terminal velocity. The terminal velocity achieved by a falling object is proportional to the ratio of its frontal surface area and mass. So, a 1-kilogram lead ball will reach a much higher terminal velocity than a 1-kilogram lead sheet, and a 1-centimetre-diameter stone will reach a higher terminal velocity than a 1-centimetre-diameter sponge.

In 1926, the British scientist J.B.S. Haldane summed up the effects of size on what happens to a falling animal in an essay entitled 'On Being the Right Size'. With typical flair, he wrote, 'You can drop a mouse down a thousand-foot mine shaft; and, on arriving at the bottom, it gets a slight shock and walks away – a rat is killed, a man is broken, a horse splashes.' It's all to do with the ratio of surface area to volume – or, put simply, the bigger they come, the harder they fall. If you want to see this effect in action, throw a garden pea and a melon up into the air over a hard surface and see what happens. I tried this demonstration in a lecture once and, having picked a melon that was rather on the large side, when it hit the stage, it splattered over most of the front row. The pea was fine, though. The relationship between gravity and the size of an animal also explains why emus and ostriches can't fly, elephants don't jump and why there is no height from which a falling insect would be harmed.

Despite the laws of physics, humans are capable of believing all sorts of ridiculous things. How many times have you read something like, 'If a flea was 15 centimetres long, it could jump over a tall building like St Paul's Cathedral'? You can't scale these things up. A 15-centimetre-long flea could jump no higher than a normal flea, and probably a bit less. A large flea would be a lot heavier and have a bigger surface area to be affected by air resistance, and much of its power would be used up simply overcoming the force of gravity. The reason fleas have a relatively impressive jump is because they store energy before they launch themselves into the air. This stored energy can be released rapidly, allowing them to accelerate their small mass in under a thousandth of a second to a speed of over 1 metre per second. Even with this mechanism, their jump only has a range of between 10 and 20 centimetres. The impressive feature of the flea catapult mechanism is that it allows these tiny insects to jump about 45 times their own length, whereas larger animals, although they can jump further, can only jump a few times their own length.

Another advantage of being a small animal like an insect is that they are relatively more powerful than large animals. The cross-sectional area of insect muscles is large compared to the mass they are supporting. I am able (or I used to be able) to carry a load about equal to my own body weight (for a small distance), whereas the load-carrying ability of an ant is many times its own body weight.

What other consequences are there to being small? The world is a much finer-grained environment to small animals and offers a lot more places to live. Insects can occupy many more ecological niches than larger organisms can. Microhabitats are important to insects for other reasons. Environmental conditions vary considerably on a small scale and, in any habitat, there will always be a range of conditions available. Cooler, moister microhabitats can be found near plants or under stones or by burrowing underground – even moving a short distance will make the hottest conditions survivable. Small animals like insects are much more able to take advantage of thermal variations in the environment. An elephant cannot move from an air temperature of 50° C to a much more bearable 30° C by simply walking around to the other side of a leaf.

Wonderful wiring

The nervous system of an insect provides information about the environment outside the body as well about what's happening inside. The central nervous system of an insect is made up of a dorsally situated brain inside the head, connected on two sides to a large ventral mass of nerve tissue called a ganglion, which lies beneath the oesophagus. From here, a paired nerve cord runs back along the underside of the body cavity, joining a series of thoracic and abdominal ganglia along the way.

The brain, ganglia, nerve cords and the major peripheral nerves are sheathed by a supportive, corset-like layer inside which a sleeve of cells, called glial cells, act as insulation. These glial cells surround and insulate nerve cells and are also tightly linked together in a layer around the brain, ganglia and main nerve branches, forming what is known as a blood–brain barrier. This means that the nervous system is completely separated from the rest of the bodily fluids and can thus function efficiently at all times. The all-important cells of the nervous system rely on precisely controlled concentrations of sodium, potassium and other ions that generate the electrical currents that carry information around the body. They need a private pool, not communal bathing.

Imagine the modern world without electrical insulation. Incredibly careful routing of bare wires would be needed to avoid circuits shorting out. If it were not for the insulating properties of the glial cells, electrical signals passing through the million or so neurons that make up an insect's nervous system would become severely degraded. While a blood–brain barrier of sorts is found in other terrestrial arthropods, such as spiders, that of the insects is highly effective indeed, and was a major factor that allowed them to successfully colonise the land in the first place.

Super senses

Like the central computer of an intelligence agency gathering masses of information from all its aerials, microphones and cameras, the insect's central nervous system receives information from a myriad of sensory receptors all over the body. With the exception of soil-living or cave-adapted species, which are often blind, the main visual sense organs of adult insects is a pair of compound eyes. These eyes are made up of individual light-receptive units called facets, and there may be anything from one or two in some ants to more than 10,000 in dragonflies. Each facet of the eye comprises a clear lens with light-sensitive cells beneath. In day-flying insects, the image received by the eye as a whole is made up of a mosaic of spots of differing light intensity from all the separate facets. The more facets the eyes have, the greater will be the insects' visual acuity. The eyes of night- or dusk-flying insects have a different internal construction and sacrifice visual acuity in favour of light-gathering power. Colour vision has been shown to occur in all orders of insects, and many can also detect the plane of polarised light, which offers important cues for navigation and telling the time.

In some insects, such as dragonflies, which are incredible aerial hunters, the eyes are so important that they become very large indeed, sometimes taking up most of the surface area of the head. The compound eyes, of the males of some mayflies can be divided into two zones. The lower zone looks like a regular compound eye, while the upper zone is like a blunt, mushroom-like stalk with all the facets facing upwards. This allows the males to spot females above them in a mating swarm. In whirligig beetles, the eyes are completely divided in two, with one part for seeing above water and the other for seeing below the water surface.

Another sort of insect that relies heavily on good eyesight, the key to their success as blood-feeders, are horse flies. I first got a real, close-up look at their feeding strategy when making a film about how horse flies see horses. I caught a horse fly in a tube, and when the macro-camera was ready to record, I upended the tube onto my bare forearm. The horse fly immediately started to slash enthusiastically at my skin with its blade-like mandibles. Horse fly eyes have inbuilt polarisation filters that make large, dark objects such as cattle and horses stand out clearly against the background. Light-coloured horses are not so obvious or easily found;

while they might get bitten less, they do suffer more from sunburn. A likely explanation of why zebras look the way they do is because it seems that blood-feeding flies don't like landing on stripes. But the jury is still out on exactly why stripes seem to disrupt their visual homing system.

Insects can see things that we cannot, and vice versa. In general, insects see better at the blue end of the spectrum than at the red end, and in some groups the range of sensitivity carries on into the ultraviolet. The ability of bees and other insects to see in ultraviolet is well known, and flowers, which have co-evolved with pollinators, often have distinctive markings called nectar guides, which are only visible to us if they are illuminated by ultraviolet light. The fact that insects cannot see well at the red end of the spectrum has been useful for biologists who want to study nocturnal insects, as they can be observed under red light without disrupting their normal behaviour patterns. It has also been shown that dung beetles use the Moon to orientate themselves and might also pick up clues from star patterns. It seems that we're not the only ones gazing up at the heavens.

Insects don't just see differently to us, they hear differently, too. The surface of the cuticle bristles with hairs of various types which are responsive to vibrations and even the gentlest puff of wind. Special hearing structures called tympanal organs may be present on various parts of the body and, depending on species, are responsive to sound frequencies ranging from less than 100 Hz (the low end of a human's hearing) to 240 kHz (a frequency higher than bats can hear). The cuticle also contains strain gauges that respond to stresses on the exoskeleton, and there are internal receptors that sense the stretching of muscles and the digestive tract. Chemical sense organs, called chemoreceptors, are present on the mouthparts, antennae, feet, external genitalia and other parts of the body. A book written in 1247 by Song Ci, a Japanese judge, doctor and scientist, tells the tale of what is believed to be the first recorded murder case solved through the application of forensic entomology. A villager was stabbed to death with a sickle used for rice harvesting, and all the villagers protested their innocence. The magistrate in charge asked all the villagers to bring their sickles to the village square and place them on the ground in front of them. It was not long before blow flies began to gather on one of the sickles. The murderer may have attempted to clean the blade, but the minute quantities of blood that remained were enough to give the game away. If you want a first-hand demonstration of how sensitive fly senses can be, put a piece of fish skin out in the sun and see how long it takes for flies to find it.

Insects can also 'smell' airborne chemicals by means of sensory structures found mainly on the antennae. The antennae of many male moths are feathery-looking and covered with receptors that alert the male to the sexual odours (pheromones) produced by the females. The receptors are so sensitive that only a few molecules of the right pheromone are needed to lure the males on their journey upwind to meet a potential mate.

Many insects have temperature and humidity sensors, and some can even detect magnetic fields. Certain jewel beetles can detect infrared radiation, which guides them to freshly burned trees where they can lay their eggs. The brilliant bodies of insects are more than well equipped to find exactly what they need and to avoid danger.

Winged wonders

The power of flight, which has enabled insects to take over the world, would never have been possible without a lightweight, rigid cuticle. While there might be arguments about exactly what sort of arthropod first walked over dry land, there is absolutely no doubt that insects were the first animals to take to the air. Insects evolved wings in the Carboniferous Period, around 300 million years ago, but there is some speculation as to how. Insects are markedly different from other flying creatures like birds and bats, which evolved wings from already existing limbs. Were insect wings entirely new structures that grew out of the thorax, or were they derived from already existing structure? It is thought that the common ancestor of crustaceans and insects had two extra basal leg segments – and the general consensus, based on morphology and genetic studies, is that these ancient lobed segments became incorporated into the side of body segments. Over time, they moved to the position wings now occupy. Evolution does not produce something completely new and ready for use, but gradually tinkers with and repurposes what already exists.

Flight allowed insects to disperse, colonise new habitats and escape from their enemies. Flight, combined with their small size, allows insects to get around easily. Aphids can get swept high into the atmosphere, where they will be carried for hundreds or thousands of miles. What may have started as a jump-and-glide response to attack could easily have become prolonged, and it is not hard to see how being airborne for longer or being further away from danger would be anything other than a major selective advantage.

Flying for small animals is relatively easy. The bigger you get, the more difficult it is to produce enough power. However, being very small can produce some different problems – including that you will need to flap your wings a lot faster. Insects like dragonflies power their flight by a series of nervous signals to the flight muscles – one set of muscles makes the wings go up and another set makes the wings go down. But there are limits to the rate at which the nervous signals can be sent. In large insects with long wings, the flight muscles can be attached directly to the bases of the wings, and the speed of nervous impulses is sufficient to flap the wings at a suitable frequency. Insects such as house flies have smaller wings, and the frequency at which they need to beat is far above the speed at which the nerve cells can fire. Instead of the flight muscles being connected directly to bases of the wings, they are attached indirectly to the highly elastic, box-like thorax. When the thorax is pulled together from top to bottom, the wings flick up. and when the thorax is pulled in from front to back, the wings go down. Another neat trick is that the muscles do not need to be fired once for every wing stroke. Instead, they contract rapidly in response to being stretched, so individual nervous impulses are not required for every muscle contraction. The rapid cyclical distortion of the thorax and the way in which the wings are hinged to the thorax generate the up-and-down stroke automatically by alternately stretching the opposing indirect flight muscles – the whole mechanism is the result of some elegant evolutionary engineering.

You may have heard the urban myth that some physicist once wrote that according to the laws of aerodynamics, bumble bees shouldn't be able to fly. Well, of course, bumble bees fly very well. So what's going on? It turns out that the conventional laws of aerodynamics that work well for aeroplanes and helicopters don't work when it when it comes to describing how insects stay airborne. The problem has been to find out the exact mechanism of how they do it. As the bees' wings beat, they generate tiny vortices along the upper surface of each wing. These vortices have been likened to miniature hurricanes, which, as we all know, have low pressure at their centres – and it's these eddies above each wing that generate the lift the bees need.

High-speed filming and computer analysis have shown that the very smallest flying insects – such as feather-winged beetles, thrips and minute parasitic wasps called fairyflies – stay airborne in a completely new way. To a very small insect, the air around them will seem relatively more viscous, but they still need to fly efficiently. To achieve this, the wings of the very smallest insects are no more than thin struts fringed with slender hairs and move in a manner not seen in larger insects, allowing the insects to fly at speeds equal to those three times their size.

When you watch a hover fly as it zips and darts about – seemingly able to fly in any direction with equal ease and hover in midair irrespective of unpredictable wind conditions, you can appreciate that they have mastered flight to an extraordinary degree. There is still much to be learned about how insects control their flight, and it is not surprising that a lot of research is being done on how best to make miniature aerial vehicles for search and surveillance missions. In the future, that humming thing above your head might not be as benign as the creature upon which it was based.

Fast breeders

If there's one thing insects excel at, it's multiplying. This, coupled with relatively short life cycles, means that insects can evolve much faster than slower-breeding species.

What would happen if insects were able to breed completely unchecked? Entomologists do like to impress people with phenomenal facts. One such calculation showed that the descendants of a single pair of house flies could, in a year, cover the surface of the Earth to a depth of 14 metres. Another even scarier estimate featured that enduring laboratory favourite, the fruit fly. One popular version goes like this: if a pair of fruit flies bred for 1 year, and all their offspring survived and continued to breed at the same maximum rate, the total number of flies produced would be 1×10^{41}, or 1 times 10 to the power 41, or 100 duodecillion. At any rate it's a seriously large number of flies which, incidentally, would make a ball of flies that would just about fit between the Earth and the Sun.

I suppose it depends on how tightly you packed them but, not wishing to be left out of all this multiplicative mayhem, I decided to do a rough calculation of my own. For a few years, I helped on a research project looking at two species of bean weevil and their parasites. These small beetles lay their eggs on the outsides of black-eyed beans. The beetle larvae hatch and bore into the bean, where they feed until they emerge as adults. I reckon that a pair of beetles could produce about 80 offspring every 3 weeks so. Theoretically, the number of beetles in the second generation would be 3,200 and, assuming there was enough food to go round, the third generation would number 128,000 beetles. By the eighteenth generation, which incidentally would only take 432 days, the beetles would number 1.4×10^{29} individuals. This number of beetles would occupy a volume equal to that of the Earth.

So why, I hear you ask, are we not up to our armpits in insects? Population explosions of insects can sometimes take place when conditions are just right but, in general, insect numbers are controlled by lots of different factors, such as adverse weather, disease, predation and food availability. However, the astonishing reproductive potential of many insects is beyond doubt, and they will multiply given half a chance.

The world is very different to an insect in comparison to the world that we inhabit. To understand how insects work, you need to think at their scale – but in the next chapter, we're going to move from thinking at the small scale of a single insect to the much larger scale of ecosystems. Insects are crucial to the world's ecosystems; they underpin them. In fact, I'd go as far as to say that insects make the world work.

Chapter 3

HOW TO BUILD
A PYRAMID

A living laboratory

Set on two hills overlooking Oxford, Wytham Woods is a special place to me. I must have made the short journey there from the Oxford University Museum of Natural History countless times over the years. A small private road leads up the hill behind Wytham village to a public car park and the locked gates leading into Wytham Woods. The 424 hectares (a little over 1,000 acres) that make up Wytham Woods is a patchwork of ancient and more modern woodland, as well as grassland and scrub. Nothing remarkable, you might think, but what makes Wytham Woods unique is that there have been more biologists observing, recording, measuring and experimenting within its boundaries than probably anywhere else on the surface of our planet. It is in every sense a living laboratory, and I imagine there isn't a single square centimetre that hasn't been used for some research project or other. Many of the architects of modern ecological science of the latter half of the twentieth century

67

carried out their pioneering work in Wytham Woods. One or two are even buried in the graveyard in Wytham village, and I reckon that wouldn't be a bad place to end up.

You would think that after many decades of intensive study, we would know every single species to be found in Wytham Woods, but you'd be wrong. Once, when teaching an undergraduate field course there, I gave two undergraduates the task of dissecting a small length of decaying birch log to see what they could find. They worked quietly away for the whole of the afternoon and then called me over to show me what they had found. There were plenty of woodlice, millipedes, some slugs and a number of insect larvae. They were just about to tidy up and pack their bags to return to Oxford on the coach when I went over again to see if they had found anything else. This time I slid the dish of alcohol containing their catch under a microscope and, to my great surprise, I saw a pale, wingless gall midge I did not recognise. On my return to the museum, I packed it up and sent it off in the post to the foremost gall midge expert in England. The answer came back in a few weeks. My students had found the male of an undescribed species of gall midge belonging to the genus *Bryomyia*. I was completely amazed. No one had ever seen one of these tiny flies before. This species was not just new to Britain, but new to the whole of science. How had it gone unrecorded for all these years – and in Wytham Woods, of all places?

I keep wondering about this little fly, discovered by chance and never seen again. Why was it so rare? Did it occur elsewhere in Europe and, in the UK, live right at the edge of the range of possible habitat that is suitable for its needs? What species preyed on it? Were there bacterial or fungal diseases that kept it rare? These questions are at the very heart of ecology, which seeks to understand the factors that regulate the abundance and distribution of species. And that's what this chapter is about.

The ecology of insects

The prefix 'eco' – as in eco-friendly, ecotourism and eco-warrior – is commonplace today and is shorthand for 'ecological'. Ecology, derived from the Ancient Greek word *oikos*, meaning 'house', was first used by a German scientist named Ernst Haeckel in 1866. It means the study of organisms, their relationships with other organisms and the physical environment. But the word 'ecological' has recently been used to mean something to do with the environment or some product that is less bad for the environment than some other product. It's all got rather vague.

Simply knowing what species live in a particular area does not make you an ecologist – you're going to have to find out a whole lot more to earn that title. You'll have to know how each species lives and how they die, what they need and what they don't need. In short, you're going to have to work out exactly how that bit of the natural world works, and insects are going occupy a lot of your time.

The ecological impact of insects is absolutely colossal. Food chains depend on them. Insects are the world's food. We depend on pollinating insects, especially bees, for perhaps as much as a third of the food we eat. As recyclers, flies and beetles devour carcasses, and clear prodigious quantities of dung from the surface of the planet every day. When you consider African wildlife and you hear the word 'herbivore', you will probably immediately think of wildebeest, elephants or zebras. It might come as a surprise that all the heaving, snorting herds of grazing ungulates are entirely 'out-munched', perhaps by a factor of 10 to 1, by billions of tiny mandibles.

And what about the meat eaters? I say meat eater, you may say lion. But again, insects eat many times more animal flesh than all the vertebrate carnivores put together. Ants alone are the major meat-eating species in any habitat, whether it's dry grasslands, steaming jungles or your own back garden. The natural world is governed and regulated by small species with six legs. These are the creatures that carry out the fundamental processes that keep the whole ecological show on the road. Without insects, some of the strangest mammals – the aye-aye, the pangolin, the giant anteater and the echidna – would never have evolved, and neither, it's fair to say, would humans.

Insects, with their vast numbers and the extensive interactions they have with plants, fungi and other animals, make up a large part of the mechanism of the natural world. Communities of species in an ecosystem can be thought of as forming a pyramid, with the broad bottom layer made up of plants and higher, progressively smaller layers made up of herbivores and then, on top of that, the carnivores. You can't have more herbivores than there are plants for them to eat, nor can there be more carnivores than there is prey. Stable pyramids are built from the bottom up. Learning about this stuff isn't just important to pass Ecology 101 at university

– it's important because we humans are a significant part of world ecology and understanding this and our impact may help keep us alive.

The naming of parts

But before you can do any ecology at all, you need to know who and what you're dealing with. Mathematicians sometimes say that mathematics is the primary science. I beg to differ. Before you can count anything, you need to know what you're counting. You need to be able to tell the difference between apples and pears or sheep and goats. I would say that if there was ever such as thing as the 'the oldest profession', then it must be taxonomy: the description, identification and classification of organisms. You might think of taxonomy as a rather dry and dusty subject, but it is the cornerstone of all scientific investigations and the first question we ask of anything we find: what is it? Knowing what species you're working on is rather important if you're going to be able to compare your own findings with anyone else's.

Biomass and energy

Once you know what it is that you're counting, you need to start … counting. To understand how ecosystems work, you must make measurements. You might measure the numbers of particular organisms. But much more useful is biomass, the total mass of biological material – plants, insects or birds, for example. You certainly need lots of plant biomass on the bottom layer of the pyramid. The energy plants trap by photosynthesis and convert into food is the start of many terrestrial food chains. A smaller biomass of herbivores occupies the next layer of the pyramid, and they consume as food a proportion of the layer below. The next layer and subsequent layers are made up of carnivores that prey on the species in the layers below. At the top of the pyramid sits a relatively small biomass of what are known appropriately as apex predators – these are the species at the top of a food chain. Intuitively, it is obvious that you cannot have too many cheetahs and too few gazelles, or too many gazelles and not much grass to feed them – it just won't work.

These layers of the pyramid are called levels. Plants occupy level one, herbivores occupy level two and carnivores occupy levels three and above. Only a certain amount of energy is available overall, and at each level, the amount of energy available gets smaller and smaller. Only 2% at most of the Sun's energy is harnessed by plants and converted into food. When this stored energy is consumed by a herbivore, only about 10% of it is used to make herbivore biomass. The rest is used to power the herbivore's metabolism and some is lost as heat. The same thing happens at the next level – only 10% of the energy gets passed up the chain each time. The result of this reduction in available energy is one of the reasons why ecosystems only have four or five levels.

Insects fill a large part of the second and third levels. As mentioned, they are both the main consumers of plants and the most prolific carnivores on Earth. Insects are devoured by many animals higher up the pyramid. Even the smallest of the British bats may consume many thousands of insects in a single night. A brood of great tits, researched extensively at Wytham Woods, consumes vast quantities of insect food before they fledge. I'd wager that all the great tits in the UK eat many billions of insects in a single year. Without insects, it would be hard to see how complex ecosystems could ever have evolved.

Beneath our feet

Most people will have seen colonies of bacteria growing in a Petri dish. The bacteria grow on a jelly-like substance called agar, which is derived from red algae. Some bacteria are fussy and need particular nutrients to be mixed in with the agar before it is poured into the dish or they will not be able to grow. What supports the pyramid of life on Earth? Water and air are vital, of course, but deciduous forests with ancient trees encrusted in lichens and mosses, tropical moist forests with orchid- and liana-laden branches, and savannahs with their vast herds of grazing animals wouldn't exist without one thing. Just like the bacteria in a Petri dish, the green plants that support much of the diversity of life on Earth also need a growing medium – soil. Soil may seem abundant, but it's not. The average thickness of the Earth's crust is somewhere between 15 and 20 kilometres, but this only accounts for less than 1% of the planet's volume. The Earth's crust can be likened to a layer of thin card stuck to the outside of a regulation football. At the same scale, the soil that covers parts of the land is nothing more than an insubstantial wash of thinned watercolour paint.

Next time you are in a woodland, a garden or a field, bend down and scrape up some soil. Examine it closely; smell it.

Just a handful of dirt? Certainly not. Soil is so important that our existence, and that of most other species, would be very uncertain without it. Soil anchors, supports and feeds plants, as well as regulates the flow and quality of fresh water. It is also rich in species and is a crucial store of carbon. But how much do we really know about it? Who do you think uttered the following words? 'We know more about the movement of celestial bodies than about the soil underfoot.' David Attenborough, perhaps? Or Rachel Carson? Franklin D. Roosevelt or Aldo Leopold? They have all said wise things about soil, but the words were written by Leonardo da Vinci in the sixteenth century – and they're as true today as they were then. For a natural resource that we could never manufacture, and that does so much for us, we pay soil little attention.

Beginning with the relentless wearing down of mountains by wind and rain, the formation of soil is driven by complex physical and chemical processes and the slow accumulation of organic material. Topsoil, the uppermost layer, usually the top 3 to 25 centimetres (5 to 10 inches), has the highest concentration of organic matter and microorganisms and is where most of the biological activity occurs. It can take anything from 500 to 1,000 years to create 2.5 centimetres (1 inch) of topsoil but, since the advent of agriculture, it has been steadily degraded and eroded. Regularly ploughed and exposed to the elements, soil is easily washed from the land and begins a one-way journey down to the sea, where it will settle and sink to the ocean floor.

Healthy soil is a hidden world of biodiversity. A teaspoonful of it may contain a billion or more bacterial cells, 200 metres of fungal threads and hundreds of tiny organisms from various different animal groups. But many of the creatures beneath our feet are not destined to live their whole lives in darkness. More than 40% of organisms in terrestrial ecosystems live part of their life cycle, typically the larval stage, in soil. Those colourful hover flies, long-legged crane flies and fast-running ground beetles you see above ground were once denizens of the dirt.

During heavy rains you will see that the water running off the land is reddish brown. All this water is heading through the waterways to the ocean, and much of it will be recycled to fall somewhere as rain. But it carries with it a precious resource, which cannot be recycled. The rate at which soil is being lost can be measured easily. All you have to do is collect some water and evaporate it. What's left is the soil that has washed from the surface of the land. World-wide, the loss is between 25 and 36 billion tonnes every year. Soil erosion ultimately leading to desertification is taking place on every continent. Industrial-scale agriculture, the business of converting global resources into human flesh, is mostly to blame. Every year more than 80 million more human beings have to be fed, but at the same time, billions of tonnes of soil are being lost. Intensive crop production leads to the loss of soil nutrients and the increasing reliance on fertilisers. Irrigation inevitably leads to the accumulation of salt deposits. Millions of hectares of once productive land have already been severely damaged in this way. In years to come, well over half of the world's agricultural land may be degraded – what will we do then, I wonder?

Our ecological pyramid sits on soil and, because of it, an astonishing change was able to occur – one that shifted the direction of life on Earth. Out of the soil, a whole new type of plant would appear, and these new plants would kick-start a biological revolution.

A world of colour

Around 100 million years ago – a mere one and a half paces from the end of the timeline I described in Chapter 1 – the appearance of flowering plants brought colour to a world of browns and greens. These plants were more efficient than the conifers and cycads that already existed. They were better at capturing the Sun's energy, they absorbed more carbon dioxide and produced more oxygen. Most importantly, their reproduction was much more efficient. Rather than relying on the wasteful and haphazard nature of wind pollination, these plants thrived by securing the services of insects. To do that they provided their prospective allies with energy-rich fuel in the form of nectar to power their flight, and lured them in with colours and odours. The collaboration between flowering plants and pollinating insects is the most wide-spread and significant symbiosis on the planet today and, as a result, flowering plants now make up 90% of all plant species and more than three quarters of all insects depend on them. Without this one type of interaction, the study of ecology would be a great deal simpler.

Many insects can act as pollinators, but it was bees, who had themselves evolved from wasp-like species, that made the switch from prey to nectar and pollen as their primary food source. Of the approximately 20,000 known species of bees, most live solitary lives – but some, such as honey bees, bumble bees and stingless bees, are social.

From the biggest (Wallace's Giant Bee, *Megachile pluto*, a rare Indonesian species around the size of a human thumb) to the smallest (a minuscule mining bee, smaller than a grain of rice, that lives in the southwestern United States), bees are, without doubt, the single most important group of insects on Earth. I am not going to list all the crops that bees pollinate, but from apples to avocados, beans to blueberries, and cucumbers to cherries, one in every third mouthful of the food we eat relies on the pollination services of bees. That's quite a thought, isn't it?

And in payment for all the things that bees do for us, we steal things from them!

Sweet rewards

Early humans collected honey from wild bees and bee-keeping is documented from as long ago as 2000 BCE. Honey bees belong to the genus *Apis*, and the best known of the seven species is the Western Honey Bee, *Apis mellifera*, which has been spread by humans worldwide. Next time you eat honey, you might remember that a lot of work has gone into making it. One jar of honey has required many millions of visits to individual flowers.

While honey bees are quite good pollinators, they are not as versatile as bumble bees. The 250 world species of bumble bees are found mainly in the northern hemisphere, although some species can be found in South America. In the UK there are about 25 species, but sadly, we have not been taking care of these national treasures. Three species of bumble bee are no longer found in the UK and the numbers of half a dozen others are declining. You don't have to look very far for the reasons. Agricultural intensification and the loss of over 97% of the UK's wildflower-rich grasslands in the last 50 years have, unsurprisingly, had a huge impact on them.

Queen of the machair

It was early July a few years ago that I boarded the ferry at Uig on the Isle of Skye bound for Lochmaddy on North Uist in the Western Isles, also called the Outer Hebrides. Although I had been on many summer holidays to the west of Scotland as a boy, I had never been to the Western Isles. I was heading for Balranald to film the Great Yellow Bumble Bee, *Bombus distinguendus*, which lives on the machair, the coastal grassy plains found on the west coast of Scotland and Ireland. This large and handsome bumble bee is now only found in the far north and west of Scotland, but 50 or 60 years ago I wouldn't have had to travel nearly so far; it could have been found in many locations scattered widely across the UK.

The soil of the machair is unique. Seashells are pounded to dust by the waves and the strong Atlantic winds carry the dust onto the land. This calcium-rich windfall, together with traditional crofting practices, leads to an explosion of wildflowers during the summer months. As I stood in the bright sunlight, scanning the flowers of bird's-foot trefoil, red clover and kidney vetch, hoping to catch a glimpse of a bee I had never seen before, I knew that this was an incredibly special place. When a queen bee eventually appeared, flying low over the machair, she was quite unmistakable. In that instant, with the noise of the wind and crashing Atlantic breakers in my ears and this beautiful queen of the machair in my hand, I could not have been any happier.

Can you imagine a summer without the buzz of bumble bees in meadows, gardens and hedgerows? These endearing and industrious insects are keystone species in our landscape. They are major pollinators of wildflowers such as foxglove, lavender, honeysuckle and bluebell, and they are of enormous commercial importance.

It is likely that every tomato you eat was pollinated by a bumble bee. This is because of a pollination technique particular to these bees. The anthers of tomatoes will only release their pollen when they are vibrated at a frequency of around 400 Hertz, and bumble bees do this by grabbing hold of the anthers and vibrating their flight muscles. Before bumble bees were reared commercially for glasshouse pollination, this 'buzz' pollination, as it is known, was done by hand with a mechanical vibrating device.

That bees are irreplaceable and what might happen if they disappeared is no longer an academic debate. In some parts of the world where pollinators have declined dramatically, farmers are already having to pollinate their fruit trees by hand. We have set in motion processes that could lead to the extinction of some of the planet's most important insects, an indispensable part of our ecological pyramid. The loss of bees could trigger a global biodiversity crisis threatening the survival of a quarter of all terrestrial species. At a time when one sixth of all human beings are hungry and there are growing concerns over food security, losing the services of bees would be catastrophic.

The sorcerer's apprentice

Unfortunately, it is often the case that we learn about the workings of the natural world only when it goes awry. There are many examples where humans have thought they understood how ecosystems work, only to find out later that fiddling with one layer of the pyramid can cause things to go horribly wrong, often with undesirable and unintended consequences. It's happened so many times over the years, I really do wonder if we are capable of learning from our past mistakes. Nothing illustrates this better than the introduction, accidentally or otherwise, of species to parts of the world where they don't belong.

Fiddling with natural communities that have evolved over a long period of time is rarely a good idea, and I want to show this by telling you the story of Flathead Lake in Montana, in the western United States. Now I know that many of you will put your hands up to point out that this story does not involve insects. It does not. But it's such a good story, and illustrates my point so well, that I just can't resist retelling it – and insects are close cousins of the main protagonist anyway.

Records from the late 1800s show that Flathead Lake was home to 10 or so native fish species, but as people moved into the area, they introduced two dozen other non-native fish species to the lake. These fish were known to be good eating, so nobody thought too much about it. A little later, Kokanee salmon were also added to the lake, and they did very well. By 1980 or so they were abundant, attracting anglers of all kinds: bears, bald eagles and humans. Wildlife enthusiasts travelled from all over to see the spectacle. A year later it was decided that things could be improved even more and an alien creature, an opossum shrimp, *Mysis relicta*, was introduced to provide even more food for the salmon. More food, more salmon, more money – you can see where this is heading. No one thought that this would be anything other than a triumph, but it was the complete opposite. The opossum shrimp ate all the zooplankton that the fish would have eaten and, worse still, they only fed at night, returning to the depths of the lake during the day when the salmon were actually feeding. By 1986 the opossum shrimp population had exploded and, shortly after that, the salmon population plummeted. With few salmon around, the bears and the bald eagles left to find food elsewhere.

I can't help but be reminded of *Fantasia*, Walt Disney's classic 1940 cartoon film based on the 1797 poem 'The Sorcerer's Apprentice' by Goethe. Mickey Mouse, the sorcerer's apprentice, gets tired of fetching water to fill up a large tub and uses a magic spell to enlist the help of a broom. He falls asleep and later wakes to find the tub and the room full of water. The enchanted and tireless broom had done its job well. In desperation, Mickey seizes an axe and reduces the animated broom to splinters. Unfortunately, each splinter becomes a new broom and they keep on fetching water. Mickey realises too late that he doesn't really know what he's doing and that his hubris has landed him in a whole heap of trouble. Only when the sorcerer returns is normality restored.

In Flathead Lake, all of this could have been avoided if they had looked at the ecology of the lake and the ecology of the alien species they were about to introduce. But making money now and worrying about the consequences later has been long been our guiding principle.

Even when we do stop to think about the consequences, we can still get it wrong.

Useful invaders?

Aphids can be really serious pests to all kinds of crops, doing damage by their feeding and also by transmitting viruses that infect plants. Happily, ladybird beetles and their larvae are well known for having a voracious appetite for aphids. One adult ladybird can munch its way through several hundred aphids in a single day and because of this, they are regarded as beneficial insects and often used as agents of biological control. Surely this is a much better idea than spraying toxic chemicals all over the place. One species, the Harlequin Ladybird, originally from East and Central Asia, was introduced to North America just before the First World War, and it worked rather well. Not only that, it did not seem to be able to survive on its own once the aphid outbreaks had been suppressed – an ideal situation, you might think. But this convenient state of affairs would eventually change, and in the last 25 years, the Harlequin Ladybird has been able to spread across the continent, becoming the most common ladybird species in the United States. Also used as a biological control measure in Europe, the species has spread widely. Well, what's wrong with that? This ladybird eats aphids, which is exactly what's required – but ecology is never simple, and it wasn't too long before problems became apparent.

PROFESSOR HELEN ROY

For the love of ladybirds

Professor Helen Roy is an insect ecologist with a passion for lady-birds. She leads the UK Ladybird Survey and writes extensively about them. I spoke to her at the UK Centre for Ecology and Hydrology. At the time, Helen was the president of the Royal Entomological Society, a post she stepped down from in 2022. I began by asking her what it was that first interested her in these rather attractive insects.

'I've always loved being outside and, in the mid 1970s, there were huge numbers of ladybirds around, mainly seven-spot ladybirds I found out later. The garden of our small, semi-detached house on the Isle of Wight was a haven for ladybirds and me. There were all life stages present, and it was absolutely fascinating. There was so much I didn't know and I was absolutely captivated.'

I remember around that time there was a massive 'outbreak' of ladybirds and the press was full of horror stories about invasions of these insects and people getting bitten by them.

'Ladybirds will bite – they have biting mouthparts and will take a tiny little nip, but it's so inconsequential that it's almost a pleasure to be bitten. The "outbreak" saw red tides of ladybirds, and the numbers were just soaring. But, of course, as a small child on the Isle of Wight, I had no idea what was going on across the country. All I knew was what was going on in my garden – and what was going on in my world was tremendously exciting.'

Professor Roy served as the president of the Royal Entomological Society for four years, only the third woman in that role.

'The society isn't far off from being 200 years old, and not many women have been presidents in that time. It's an incredible privilege. If you'd told me during my childhood and teenage years, when I was out doing all the natural history I could possibly do, that one day I would be the president of the Royal Entomological Society, I wouldn't have believed it. I would have thought, "What an incredible dream to have."'

There are currently 47 species of ladybird in the UK, with new arrivals being discovered all the time. They're generally thought of as a good thing; if one had to choose a gardeners' friend, it would be the ladybird, because the larvae and the adults have a voracious appetite for things that gardeners don't generally want. Professor Roy explains:

'So many of them are predators of aphids and scale insects, things that gardeners don't want to have feeding on their roses, broad beans or whatever else. An adult ladybird can eat about 60 aphids in a day and a larva can do something similar. When you think about the number of larvae that are around in the summer months, it's a spectacular team of insects who are doing this fantastic work. Some species are mildew feeders and a few feed on plants as well. On the whole, certainly in the UK, there are no species that are thought to be a pest to garden plants. However, the Harlequin Ladybird has turned things around a little bit.

'The harlequin is about 6–8 millimetres in length, so as ladybirds go, it's quite large, and it has a very wide diet. It feeds fantastically on pest insects, but it will also eat the other things that also feed on the pest insects. It will eat whatever it bumps into! The concern is that, now, in an urban environment around 80–90% of the ladybirds are Harlequin Ladybirds. That's quite a shift. That could be great from an agricultural perspective, because if this Harlequin Ladybird is such an effective predator, it's going to do brilliantly at eating all of those pest insects. But of course, in ecology, balance is important.

'When we think about all of the different species of ladybirds that feed on pest insects, each one does things a little differently. There are some ladybirds that wake up from winter very early on; some wake up later. Some are active earlier on in the day; some are active in the middle of the day. Some of them like to feed on trees; some like to feed on nettles. Some like to feed at the top of the plant; some in the middle of a plant. There's all kinds of differences among them. When they're all there together, they form the pieces of a jigsaw. If you only have one piece – one species – what does that mean in terms of the resilience of the system as a whole? And our predictions would be that it won't be such a resilient system.'

Because habitat is being lost so fast, we are probably losing species faster than we could ever name them. And the thought that we

are now saying things like: 'How many species do we need? Do
we need all of them?' is shocking. Professor Roy agrees.

'I don't want a world where I'm not seeing all of the species
of ladybirds that I have seen before. Future generations
should also be able to see them.

    'There is no doubt that each species plays a slightly
different role within quite complex ecosystems. And as our
understanding of the resilience of an ecosystem is limited,
we don't know where the tipping points are. Such that if
the ecosystem shifts – to where one species is now extinct
– we're never going to be able to get back to that previous
system. And we have no knowledge of whether that new
system works, or if we should we be worried or concerned.
I think we have many, many unanswered questions.'

The Harlequin Ladybird is a case in point. It might even be
threatening other (useful) species of insects.

'We've done a lot of experiments and field studies looking
at what exactly the Harlequin Ladybird eats, and found it is
eating a variety of things. It is a fantastic aphid predator but
it is also eating other species of ladybirds – lacewings, hover
fly, larvae, parasitic wasps – all kinds of things. I think many
of these species will still thrive because they have other
places that they live and can escape to.

    'One of the species that we're particularly concerned
about is the Two-spot Ladybird. When I was a child,
two-spot ladybirds were incredibly common. I remember

seeing them all the time. Ladybirds have a chemical cock-tail which makes them taste really horrible, and each species has a slightly different chemical makeup. The two-spot ladybird is not very well chemically defended, so it's one of the better-tasting ones. The Harlequin Ladybird can eat it with no problem at all. There's also a twist in terms of its development times, such that when the Harlequin Lady-birds are large and hungry larvae, the two-spot ladybirds are often just sitting around as pupae [inactive before adult stage] and, unfortunately, the Harlequin Ladybird larvae eat them. Not only that, the harlequins outcompete the little two-spot ladybirds when they are adult because they're much more effective at feeding on aphids. Our analysis is that we're seeing quite a dramatic decline in two-spot lady-birds as a result of the harlequin.'

There is a belief that the more species you have in a habitat – the more complex and diverse an ecosystem is – the more resilient it is. Perhaps that is one of the problems we now face with the damaging effects of global warming and introduction of alien species. The more species we lose, the less resilient these habitats will become; the more fragile that whole system becomes. Profes-sor Roy elaborates:

'That hypothesis concerns us a great deal. We need to gain a better understanding about the ways in which networks of species interact and how many linkages you can lose before the system falls apart. It would be wonderful to think we could maintain the interactions that exist now, but we have to consider the alternative because the world around us is changing so very rapidly.

'We need to take action now. What's really encouraging is that there are things that we can all do to make a difference. It's wonderful to see the ways in which people are changing their behaviours, or looking after the green spaces where they live.

'It's like the pieces within a jigsaw. As you take them away, you suddenly lose the picture. Even in habitats that may be considered simple – a field of wheat for example – there are creatures that are living on the ground, in the soil and in the air above the wheat, it's remarkable. Then consider the complexities within an ancient woodland – there'll be even more species, even more interactions. If we remove some of the species, that will have consequences throughout that whole network. Some changes we can predict, some we can understand, and some, at this particular moment, we don't understand. That's why there remain big questions around stability, the resilience of systems, and why we need to get better at understanding them.'

While it's true that many of us are very concerned about the changes in our ecosystem, there are a huge number of people who regard insects with some trepidation and don't necessarily want to interact with them – or actively help them.

'I was chatting with others at the Royal Entomological Society about whether people's attitudes towards insects have changed – there's not a lot of evidence for it, but we *can* see an increase in the number of people getting involved with biological recording, which I take as an encouraging sign. Many of those people are also doing things in their gardens or in their local habitats. So there is a shift happening. Is it enough? I don't know. I think we've just got to show people how magical the world of entomology is. There's nothing like it in terms of the diversity of life histories. I always believe that when people hear about these amazing creatures they will be inspired, and I think that's what we have to work towards. We need to tell the stories of the wasps on their behalf, and share the excitement and the wonder of what those wasps are doing on *our* behalf.

'We are most definitely not separate from the natural world. By recognising the connection that we have with it we can discover the beauty in our interactions with that world. It is absolutely critical that we think in these terms, because we are so reliant on the natural world, in terms of the food that we eat, the quality of water, the air we breathe, all of it. We are reliant on the species that help us along the way: the decomposers, the pest controllers, the pollinators and, of course, many of them are insects. It's crucial that people recognise that connection.

'Going forwards, I hope that people will share the joy of the natural world, have a much better understanding of the role that insects contribute, and therefore feel a sense of motivation to look after them. I hope that humankind discovers some amazing solutions on a very large scale, so that insects can thrive alongside us.'

Professor Roy believes we all have a part to play and shares some practical advice on maintaining the ecosystems around us.

'We need to take responsibility and look after the planet that we live on. A few years ago, the Intergovernmental Science-Policy Platform on Biodiversity and Ecosystem Services published a global assessment with stark messages, like "one million species are on the verge of extinction", and emphasised the need for transformative change. We need to consider what that "transformative change" is exactly.

'This action can't be just greenwashing – we need things that will make a big difference. We need to educate people so that they really do understand their own reliance on the natural world – that so many foods would disappear if we didn't have pollinating insects or that so many of these insects are involved in decomposition processes.

'Start by trying to think like some of the insects using your garden: what is it that they're going to need and when will they need it? For example, try and have some flowers that are in bloom throughout the year. Leave the dandelions or patches of clover in your lawn, but have some bare patches as well. Create a variety of habitats within your own garden. It doesn't matter how big or how small your space is, everyone can do this. When you think about the size of an insect, you can create a lot of little microhabitats even in the smallest of gardens.

'Choose to grow simple flowers in a variety of different colours, because some insects prefer certain colours over others. Honeysuckle is amazing for some insects.

'Consider not sweeping up all of the leaves in autumn. I don't understand why people go around with leaf blowers and remove all the fallen leaves, because those leaves can not only provide food for insects, but also a habitat for the winter months. For example, different ladybird species like different places to spend the winter. Seven-spot ladybirds love to go underneath the leaf litter and spend their winter under the mulch. Others will go into the little cracks and crevices of sheds, trees or fences. Leave a variety of different places for those insects over the winter months.

'Don't try and destroy all of the pest insects on the vegetables that you might be growing. Allow them to be there for the insects that will be feeding on them.

'For people who live in a flat and have a small window sill, plant flowers that are attractive to pollinating insects. You'd be amazed at what comes to visit, and what's occurring within the soil of that window box.'

Helen is right – we need to act now. We can no longer afford to lose any more habitats – the rainforest in particular, which now only covers less than 6% of land surface area.

'We need to protect the habitats that we have, and restore those that we have damaged. Note that restoration doesn't mean recreating exactly what was there before, but building novel and innovative solutions, so that we can support as many of the species and their connections with one another as we possibly can. That's where the climate-change scientists, the invasion ecologists, the people interested in the study of habitats, have all got to come together and sort it out.'

Friend or foe?

The harlequin eats aphids – fine – but it eats so many of them that the food for other ladybird species is reduced, and when aphids aren't around, it also eats other ladybird larvae as well. Depending on your point of view, the Harlequin Ladybird is either an incredibly useful thing to have around or it is a serious threat to the survival of other ladybirds.

Things may yet settle down into some sort of equilibrium, but only time will tell. One little ray of hope in the Pandora's Box we have opened is that the defensive chemicals in the yellowish fluid these ladybirds secrete from their leg joints in response to attack have been shown to have useful antibacterial properties and can even slow the growth of the protozoan responsible for the most serious form of malaria.

As you might expect, the sorts of species that are good at invading new territory successfully are capable of breeding fast, have the ability to disperse easily, are tolerant of a wide range of environmental conditions and have broad tastes when it comes to food. So it's no surprise that insects fit the bill rather well – and many have become cosmopolitan in distribution. Several species of beetles, bugs, flies, cockroaches and wasps have spread far from their home range, typically by aeroplane or ship. Imagine trying to make sure that every piece of cargo, every morsel of food and every single fruit is free from insects. It's an impossible task. Mostly, things work out OK, but every now and then something gets through. When this happens it's not always a problem because the chances of any one stowaway surviving are pretty low. But if it keeps on happening, sooner or later, one of them is going to make it and, in some cases, make it big.

There is still a prevalent attitude that animals and plants exist solely for our benefit or, at the very least, should not cause us any problems or inconvenience. When no obvious benefit is apparent, some people begin to wonder if they are any 'use' at all.

What's the point of a wasp?

I often get asked, 'What is the point of a wasp?' I assume that the questioner means the black and yellow striped social wasps that buzz around picnic tables in the summer, as opposed to the many thousands of species of small parasitic wasps whose activities go largely unseen. Of course, it all depends what is meant by 'point' – do they mean what use are they and to whom? I usually say that the 'point' of a wasp or a fly or indeed of a human being, biologically speaking at any rate, is to make more wasps, flies or human beings.

But yellow jackets and hornets are common and can provoke strong reactions. Many people do not like them at all, and pest controllers make a great deal of money from exterminating them, but these wasps are incredibly useful as 'pest' controllers themselves. Unlike solitary wasps, which provide their young with all the food they will ever need to reach adulthood with a well-stocked cell, yellow jackets constantly bring food back to the safety of a well-defended nest to feed their developing brood as it grows.

Their nests are composed of a papier mâché-like material made from chewed wood fibres that the wasps mix with their saliva. The nests may be located underground, in outbuildings and roof spaces or among vegetation.

The typical yearly cycle begins when rising spring temperatures wake a fertilised queen from her long months of hibernation. Her first job is to find a suitable nest site and make a small umbrella-shaped nest suspended from a short stalk. This primordial nest may comprise 25 open, downward-facing cells. The queen rears her first few offspring, feeding them with a mushy mass of chewed-up insects. A few weeks later with new workers to forage, feed their sisters and build more cells, the queen can devote herself to egg-laying full-time. The worker wasps, which are smaller than the queen, are sterile. By the middle of summer, the 'wasp factory' is in full swing. More horizontal paper combs, separated by tough stalks, have been added and the whole nest is enveloped in many papery layers.

Starting at first light and working through until darkness falls, many hundreds of wasps per hour stream out of the nest entrance to go hunting for prey. If the prey is small, they will bring it back whole or, if larger, it will be butchered on the ground. The workers reduce the mass of the prey by chewing off heads, legs and wings, and only carry the high-value meaty bits back home. The foraging workers are able to overpower quite large prey and will also scavenge meat from dead animals. The population can increase rapidly and a nest can reach an enormous size and contain many thousands of workers.

Towards the end of summer, the queen will lay eggs destined to become either males or new queens. Eggs that will give rise to queens are deposited in special, larger cells and are treated differently to other larvae. It's about now that the worker wasps begin to be attracted to sweet liquids such as honeydew, the excrement of sap-sucking insects such as aphids. Their interest in jam and soft drinks often brings them into conflict with picnicking humans. Now the old queen dies and nest-building stops. In late September the new queens and males leave the nest and mate (each queen will mate with more than one male). The males die and the fertilised queens must find a sheltered place to hibernate. The first frosts signal the death of the colony, which has already been in a steady decline. All the workers will die off as the food supply disappears and temperatures drop even further. In countries without a cold winter, the nests simply keep on growing, some attaining a very great size indeed.

Social wasps are simply one part of the ecological pyramid. They have evolved to take advantage of a food source and can regulate the numbers of insects whose populations can increase rapidly.

In any case, if you're a gardener, you should be delighted if a queen wasp decides to build a nest near your patch.

Knock-on effects

But what about insects, such as mosquitoes, that really do cause immense human suffering? Do we really have to put up with them on ecological grounds? Of course, it would be difficult to exterminate all mosquitoes, and it's only some mosquitoes that are disease vectors. The majority of mosquitoes are part of a complex ecological food web, and their wholesale removal would have serious and unpredictable consequences for many aquatic and terrestrial species.

Yet one more classic example of the far-reaching damage that can be done through a lack of understanding of how ecosystems work was the Four Pests Campaign in China during the Great Leap Forward that took place between 1958 and 1962. To improve life for everyone, it was decided that four sorts of animal – flies (presumably those deemed a nuisance, such as house flies and blow flies), mosquitoes (also flies of course), rats and sparrows – were to be exterminated. Rats and flies proved too much to deal with, but authorities did make serious inroads into populations of the sparrow. It was thought that the birds ate huge amounts of rice and grain and every measure to deter and kill them was put into practice. The sparrow-killing went on for two years, until a Chinese ornithologist pointed out that rice yields were actually dropping, not increasing, as had been the plan. This, he showed, was due to the fact that sparrows also eat a lot of insects. Grasshoppers and other crop-devouring species had increased dramatically, and starvation for millions was unavoidable. In the end, thousands of sparrows had to be imported and, in 1960, Chairman Mao stopped the catastrophically useless campaign against sparrows and turned his attention to bed bugs instead. At least the eradication of bed bugs was unlikely to cause much in the way of an ecological imbalance.

But we should not feel too smug, because large parts of Britain's landscape are completely man-made. The picture-postcard image of rural beauty is largely manufactured. Woodland has been felled, land has been drained, streams and rivers have been diverted and grasslands have been artificially fertilised to make better fodder for grazing animals. But even if we did nothing to a habitat at all, it would still gradually change over time. Ecologists call this change succession, and it is an entirely natural process that takes place over decades or centuries.

Have you ever seen a long fence running across the land where the vegetation on one side looks very different to the vegetation on the other side? Or perhaps you've seen a small island in the middle of a lake that looks positively overgrown in comparison to the land around the lake? Perhaps the farmer or landowner has done something to the land on one side of the fence but not the other? Perhaps the small island has been planted up to make it more attractive? In fact, the explanation is grazing pressure, it's just that you don't necessarily see it happening. A great experiment that can easily be done is to set up an 'exclosure'. The idea is to keep vertebrate herbivores such as rabbits and deer out. Erect a wire mesh fence of sufficient height and continue the mesh down well below soil level, turning it horizontally facing out to keep rabbits from digging underneath. Then, sit back and see what happens. In a woodland setting the plot will become crowded with plants and tree seedlings will begin to sprout in profusion. What has been keeping all of this greenery at bay has been constant and relentless nibbling. The often bleak-looking and denuded Scottish Highlands would not look the way they do today if it were not for sheep and deer, not to mention the heather-burning that allows large numbers of imported birds to be reared simply for the 'sport' of shooting them.

A good example of what happens when grazing pressure is suddenly reduced can be seen at Wytham Woods among many other sites. As a result of myxomatosis, a viral disease introduced in the 1950s to control burgeoning rabbit populations, grasslands became overgrown and gradually changed into thorny scrub. The lack of rabbits had a knock-on effect on rabbit predators such as weasels and stoats, and, like ripples in a pond, the changes were felt throughout the layers of the ecological pyramid. The change from grassland to scrub would have caused a drop in insect diversity too, but eventually some tree saplings that had been spared broke out above the scrub canopy and shaded it out. Given time, the scrub will become a woodland, which will see a rise in the diversity of insects once more.

If you want a high diversity of insect species – and, as I hope this chapter has showed you, we really, really do – you must ensure that there is a patchwork-like mosaic of different habitats and microhabitats available, and you will only know what these are by studying the specific requirements of the species you want to conserve. We need to be aware that what is a brilliant location for a nature reserve for some insect or other today might not be so in 50 years' time. Climate change will see to that. But it's not a hopeless task. Simple changes in management can bring huge rewards. Simply leaving dead wood lying around is no longer seen as untidy. Woodlands are not parks after all. This has been extremely good for insects that live in decaying wood and the species that feed on them.

If I was to impart one single ecological idea to you, it would be that the ecosystems that are suitable for a rich diversity of insects to exist will be absolutely ideal for most other species as well, ourselves included. Like the sorcerer's apprentice, we need to stop interfering in things we don't really understand.

I'm going to leave the last words in this chapter to Aldo Leopold, the pioneering American ecologist and conservationist who wrote these much-paraphrased lines over 70 years ago:

The last word in ignorance is the man who says of an animal or plant, 'What good is it?' If the land mechanism as a whole is good, then every part is good, whether we understand it or not. If the biota, in the course of aeons, has built something we like but do not understand, then who but a fool would discard seemingly useless parts? To keep every cog and wheel is the first precaution of intelligent tinkering.

Chapter 4

CLOSE ENCOUNTERS AND CURIOUS COUPLINGS

Dangerous liaisons

I fell in love with the Harlequin Beetle the minute I saw a drawer full of them in the entomological collections at Oxford. This unmistakable longhorn beetle, which lives in the tropical forests of South America, is over 8 centimetres long (3 inches) and gaudily patterned with bright coral pink, black and silvery white. It is a stunning-looking beetle, but what makes it truly remarkable are the front legs, especially those of the males, which are much longer than the beetle itself. Why did evolution give this insect such ludicrously long legs? The first Europeans to see it thought they had the answer. Swayed by the length of the legs and the fact that the ends of the legs are rather hooked, it seemed to them that the beetles probably moved about the forest by swinging from branch to branch, like miniature gibbons. Of course, no one had ever seen them doing this and besides, they have perfectly good wings for getting around. No, the real purpose of the harlequin beetle's legs has to do with mate-guarding. Rather than remaining physically locked into his partner by his genitalia, his long legs are used as a corral to protect the female from the attentions of other males until she lays her eggs.

After years of admiring them in museum drawers, I first got to see a living Harlequin Beetle in Costa Rica, and I can assure you it was the most wonderful experience. A friend who met me at the airport had found one earlier that day, and brought it with her to the airport as a surprise. After a tedious long-haul flight, I really did not expect to come face to face with a species I had longed to see alive.

When it comes to reproduction, insects are in a class of their own. It is their fantastic fecundity that gives them a head start over other animals. The sexual habits and evolutionary innovations, like the harlequin beetle's long legs, can be quite an eye-opener.

Who hasn't heard the story about the female praying mantis who chews the male's head off and devours him alive while they mate? Although cannibalism like this can and does happen in certain situations, it doesn't happen that often in the wild. Smart males jump off and run away … if they're quick enough. But eating her mate is not such a bad idea. It matters little to the female if her suitor loses his head because, thanks to reflexes in his central nervous system, the half-devoured male will continue to copulate quite effectively for some time after decapitation, ensuring the transfer of his sperm.

Prepare to enter a world of femmes fatales and new men, a world of sex reversal and bondage, aphrodisiacs and orgies;

a world where titillation and trauma go hand in hand. I hope I've got your attention now. We might like to think we're pretty adventurous when it comes to copulation, but as far as insects and sex are concerned, very little is off limits. You'd better brace yourself for some bizarre behaviour.

Back to basics

To really understand what's happening when beetles bonk or flies fornicate, you need to know a little bit about their reproductive plumbing. The basic system of insects starts with a pair of ovaries in the female and a pair of testes in the male. Each of the ovaries is joined to a tube call the oviduct, which leads to the single vagina or genital chamber. In the male, each testis is joined to a common ejaculatory duct, which takes the sperm to the penis. Yes, male insects have a penis. When mating, males use their penis to fill a sac-like storage organ inside the female called the spermatheca, which branches off the oviduct, with sperm. As eggs pass down the oviduct, sperm stored in the spermatheca is used to fertilise them as they pass by. As sperm must be used on a last-in, first-out basis, it follows that the sperm from earlier matings will be right at the back end of the spermatheca. Sperm deposited by the last male to mate will be first in the queue, and has therefore got the best chance of fertilising the majority of the eggs.

No male wants his sperm to be at the back of the queue. So the need to make sure that no other males come anywhere near the female until her eggs are laid is a common feature of insect sex and has resulted in the evolution of a whole range of mechanisms, some physical, some behavioural, to ensure that this is the case. The simplest way to do this is to just not let go of the female until the eggs have been laid. By that, I mean they remain locked in position with their genitalia coupled tightly together for as long as possible. This can go on for many hours, even days. The record for this feat – more than 70 days, apparently – is held by a stick insect. Even so, the males, engaged as they are, may have to fight off other males who try to dislodge them. As a fraction of the insect's life, it would be like us having to remain coupled together for several months at a time, which would be extremely inconvenient, make driving dangerous and would probably become rather boring.

Some people think that entomologists have an unhealthy and unnatural interest in sex and, to a certain extent, this is true. But there are some good reasons for our obsession. Insect genitalia are important because differences in their structure are a sure-fire way of showing that species are different. Genitalia, especially those of males, have been sculpted by evolution to make sure that males and females of the same species stick to mating with each other, and do not waste their time trying to mate with a partner of the wrong species. Although two species may look superficially very similar indeed, a close examination of their genitals will reveal whether they are different species or not. Evolution also shapes insects' sexual behaviour so that the courtship display of one closely related species will simply not cut the mustard with the wrong female, and may even result in injury or death.

The eyes have it

When you see an odd bit of insect morphology – a weird structure that doesn't seem to have any obvious function, nine times out of ten, it will be to do with sex. A great example of this are stalk-eyed flies. The first time I saw one of these odd flies was in a rainforest in Southeast Asia. Being there, gazing at it, gave me one of those wonderful moments of meditation, when nothing else in the world seemed to matter. I thought about catching it as a specimen for the collections in Oxford, but somehow, I just couldn't bring myself to do it.

The fly had the strangest-looking head of any insect I had ever seen. Its eyes and antennae were set far apart on long stalks that stuck out sideways from its head, and I wondered how it saw the world. It would be hard to sneak up on this fly. The span across its eyestalks was greater than the length of the entire fly, which probably meant it had near 360-degree vision. The eyestalks of female stalked-eyed flies are not nearly so big, so telling the sexes apart in the field is quite easy. The question is, why do these flies look the way they do? Of course, it's all to do with sex. Females need to be able to judge the quality of a potential mate, and in these flies, females are more attracted to males with longer eyestalks. The best explanation is that having long eyestalks is costly and the most well-fed males will have the longest eyestalks. This character has been selected by the females over time as a signal of fitness.

Just like stags, male stalk-eyed flies fight over patches of territory that will give them access to females. Males with longer eyestalks are more aggressive and win these contests far more often than males with shorter eyestalks.

As they face up to each other across a leaf, it must be pretty clear to both parties whose eyes are further apart. Why, if it gives well-endowed males such a sexual advantage, do their eyestalks not just keep on getting longer and longer? There must come a point surely when ridiculously long eyestalks simply get in the way or cannot develop properly in any case. At that point, the sexual selection that drove their elongation is opposed by natural selection encouraging smaller, more manageable eyestalks, and a sort of stalemate is reached.

Meeting up

But how do insects get together in the first place? Evolution has gifted insects a whole range of techniques to attract each other or, at the very least, ensure that the two sexes are in the right place at the right time. After all, there's not much point in bumbling about aimlessly hoping to bump into the right mate – although it might be what many of us do. For many insects, the process of finding a suitable mate and keeping them around long enough to have sex can be a complicated and exhausting process.

When I was a child, on the day of the Sunday school summer picnic, we all went by bus to a farmer's field just outside Edinburgh. For some reason, which still is a bit of a mystery, the required dress for children mucking about in a field was white – white shirts, white shorts, white socks and white gym shoes. Sartorial insanity, of course, and within half an hour of arriving, I had proved this to be the case by slipping and falling backwards into a very fresh cow pat of considerable size. The usual 'handkerchief and a bit of spit' technique was not going to be sufficient this time. I needed industrial cleansing and my beleaguered mum spent the rest of the morning trying in vain to make me respectable.

As an adult, I find cow pats even more irresistible. Not long ago, I sat for an hour watching the flies that had gathered on some fresh cow dung. From all the insect activity I could see, it was clear that the animal that deposited the dung had not been treated with the drugs that farmers often use to promote growth and treat worms. I could even see dung beetles popping in and out of holes they had made in the sloppy mass, their wing cases shining in the morning sun. The prophylactic misuse of veterinary drugs, especially antibiotics, is widespread, and is contributing to the steep rise in bacterial resistance to drugs. It's all a bit of a predictable mess, really – and should have been avoided. But perhaps more importantly, the whole community of insects that depend on dung is being needlessly destroyed.

The Yellow Dung Fly, *Scathophaga stercoraria*, is a common species across the northern hemisphere, and while the adults feed on smaller insects they catch, as well as nectar, their larvae feed exclusively on the dung of large animals such as cattle and horses. This recycling service is vital, as the world's cattle alone produce over 30 billion kilograms of dung every single day. This stuff really needs to be recycled.

The larger male dung flies hang about fresh dung, waiting for smaller flies to eat and female dung flies to mate with. For them, a fresh cow pat is a one-stop shop. As there are always many more males than females and as both sexes may have multiple mates, their sex life involves a certain degree of violence and coercion. If you watch the small females being summarily grabbed and forced face-down into the dung, you might imagine that mating is a male free-for-all. Indeed, the females are sometimes drowned in the dung, but after mating, it is the females themselves who move sperm into their sperm store so they might have some degree of control over the whole process. The more you watch insect behaviour in the wild, the more you become fascinated by their little lives.

Not far from where the yellow dung flies were busy making more yellow dung flies, a number of much smaller flies known as black scavenger flies were gathering on the surface of the pat. These flies typically like the dung pat to be freshly laid, as this one certainly was. Males display by flicking their black-tipped wings. They are not that fussy and will attempt to have sex with the first female they see. The female may already have been fertilised and will be trying to lay her eggs in the dung when she is grabbed again, but the male does not seem to care. He wraps his front legs tightly round her waist and will not let go, although she may try to shake him off. If he can induce her to have sex again, the next eggs she lays will be fertilised by his sperm. As I watched, I saw something I had not seen before. The pair of flies in front of me were busy copulating. To allow the female to use her ovipositor to lay eggs, the male had to disengage his genitals from her, but once she had withdrawn the end of her abdomen from the dung, he copulated with her again. This happened three or four times before they parted company. I was happy to learn a little more about their sex life, and I hope they were happy too.

Like teenagers going clubbing, insects are attracted to a favourable location, such as a cow pat, that will maximise their chances of finding a mate. But what if this sort of behaviour might expose you to danger? What if staying out of sight was the safer option?

Smells and songs

The sexual odours, or pheromones, of butterflies and moths are well known. These pheromones are extremely potent and only a few molecules are required to reach the antennae of the male moth to spur it into action. He will fly upwind, following the scent until he reaches the source. A friend of mine ordered some Emperor Moth sex pheromone to attract male emperor moths on Ilkley Moor. The first day she used the lure, a small rubber plug impregnated with pheromone, she successfully attracted half a dozen males in the space of 10 minutes. Feeling rather guilty about deceiving these wretched males to fly perhaps more than a mile for no good reason, she decided not to use it again. However, several days later, she returned to the moor without the lure but wearing the same jacket in which she had carried it earlier. Incredibly, the males, attracted to what must have just a few molecules of the pheromone that still lingered, came flying in nonetheless.

Using sound can also be a highly effective mate-luring strategy. The Deathwatch Beetle gets its name from the sexual signals that both sexes of this species make in the late spring. These beetles lay their eggs in hardwoods like oak, especially when the wood has already been attacked by fungi. Bracing themselves with their legs, they tap their heads against the wood to attract mates. Males are able to orientate themselves to females by tapping and waiting for a reply before moving off in the direction of the vibrations. Only females will answer his call. The origin of the common name derives from the fact that the faint sounds of the beetles can be heard in a quiet room or church where the dead are laid out. But this head-banging hardly constitutes a song. More sonorous serenades are employed by crickets and grasshoppers.

Male crickets produce their songs by rubbing their front wings over each other, while grasshoppers rub a row of pegs, located on the thighs of their hind legs, over the stiffened edge of the front wings.

But there is a type of cricket that takes singing, or rather its amplification, to a higher level of sophistication. Mole crickets look just like miniature, mammalian moles. Any animal that lives all its life underground is likely to evolve a robust, cylindrical shape. The front legs are going to have to be short and strong, with stout teeth or claws for digging. But, for my money, mole crickets are far more interesting because they are exceptional acoustic engineers. They construct elaborate burrows to make their songs more audible. The part of the burrow where they sit and sing is a chamber tuned to resonate at the carrier frequency of their song (about 3.4 kHz) and, from there, two exponentially flared horns amplify and carry the song up to the surface. The singing burrow is incredibly effective and, at 1 metre above the ground, the peak sound level can reach 92 decibels – enough to carry nearly a mile in quiet conditions. No wonder, then, that early manufacturers of loudspeakers copied the exact shape of the crickets' acoustic horns.

Another well-known group of insect singers are the cicadas. These insects have long life cycles; the nymphs spend many years feeding underground on the not very nutritious sap of tree roots. Cicadas often emerge as adults en masse, and this ensures that the local predators are completely swamped with food. Cicadas can be hard to spot as they are well camouflaged, but you can't miss their songs, which range from being quite light and pleasing to a cacophony of noise resembling pulsing chainsaws or mechanical grinders. The songs, which are produced by males to attract females, originate from a pair of organs called tymbals, located on either side of the abdomen. These organs work like a child's clicker toy, which, as it buckles, makes a loud click. The tymbal has a number of stiff ribs, so as the tymbal muscle beneath it contracts, the tymbal organ makes a series of clicks as each rib is deformed. When the tymbal muscle relaxes, the tymbal pops back into its original position, producing even more clicks. The abdomen of the male is largely hollow and acts like the sound box of a violin to amplify and radiate the sound. Both sexes have hearing organs called tympanic membranes on the underside of the abdomen, but the males can turn their 'ears' off when they are calling so that their hearing will not be damaged. The loudest cicada calls have been recorded at 120 decibels, which is louder than a live rock concert and well above the human pain threshold. Different species of cicada call at different times of the day. Once, when I was filming in Southeast Asia, the end of the day was marked by the discordant mass calling of a species that quickly became known as the 'gin and tonic cicada'.

PROFESSOR KARIM VAHED

The surprising sex lives of crickets

When you hear the pleasant chirruping of insects in the heat of summer, you might be surprised to learn that it has to do with sex. I met Professor Karim Vahed more than 20 years ago, when we acted as examiners for a doctoral thesis. He is particularly interested in crickets and knows more about their mating habits than anyone else I know. In a recent conversation, he described to me a basic cricket amorous encounter.

'In crickets, the males generally begin singing by rubbing their forewings together. On one forewing there is a structure that rubs over a file on the other forewing, and it's a bit like rubbing your fingernail along the teeth of a comb. The wings are rubbed repeatedly to produce a sound. If a female is attracted to the sound, she hones in on it and actually mounts the male. The male backs under the female.

'In field crickets, the male will actually switch his song when the female gets close. They have a calling song to attract the female, and then a close-range courtship song, which is quite different – a much softer, ticking sound. Again, the female will mount the male, who will transfer his sperm in a packet.

'In most cases, it's only the males that sing in order to advertise their presence and readiness to mate with the females, but there are a few species of bush crickets where the female actually has a little song to respond to the male, to say, "Here I am!"

'Unlike many species, the male doesn't have an organ that he inserts into the female. Instead, he inserts a ball of sperm, which contains a little tube that feeds up into the female. And then, in many species, he simply drops away from the female, and the transfer of the sperm into the female actually occurs *after* the pair have separated. The sperm is slowly drained out of this packet, called a spermatophore, and into the female.

'Interestingly, female choice is a really important driving force in the evolution of male mating behaviour across the animal kingdom. In some cases, the female cricket will actually choose the mate before mating, perhaps by distinguishing between each male song. We found that the song of males varies in relation to their size, age and potentially the male's condition. Females can sometimes tell a lot just from the male's calling song.

'In some cases, especially in field crickets, males are very aggressive towards one another and will have an aggressive song that they direct against rival males. Sometimes, males will go into head-to-head battles, especially if they're of similar size. One of them will open its mandibles (its jaws) and lunge at the other, sometimes even flipping them over backwards. This is so common that in China there's been a long tradition of using crickets in fighting, and putting bets on the outcome of the fight. Apparently, prize-winning crickets can exchange hands for large sums of money.'

I wondered whether the female would be fooled into thinking she had found a good mate if she was presented with a tiny male but played the song of a big male.

'It's quite possible that the female might be tricked into mating with the male. Although, when she gets up close, there's probably further assessments that happen there, as the females can then smell the male. We know that pheromones exist on the outside of the insect's exoskeleton, and these can have a role in male recognition and female choice.

'When the female meets with the male, she has another strategy up her sleeve, which she can use to control whether she uses his sperm for fertilisation. This is called "cryptic female choice" – 'cryptic' because it's difficult to detect. The female, by interfering with how long the sperm package is attached for, can actually choose how much of that sperm she takes up into her body before eating it. A lot of studies have shown that females will discriminate against smaller males using this method.

'The sperm packet itself is quite edible and can be regarded as a nuptial gift. The surrounding of the sperm packet is made of proteinaceous material. It's known that female crickets are quite promiscuous and will mate with a large number of males – it might actually be in the female's interest to mate with lots of different males, get the benefit of their food (transferred in the sperm packet), and then make decisions by the process of cryptic female choice whether they are the right mate for them.'

So, the hapless male doesn't actually know if his sperm is used or not. I wondered whether there is any way that the male cricket can make sure that his sperm is used by the female.

'In response to this very strong selective pressure, male crickets have evolved a huge variety of different strategies to try and ensure that their sperm is used by the female. One of the strategies I've studied is courtship feeding in the form of a big bowl of jelly that the male will secrete and attach to the sperm packet. Now, the female has to eat her way through this jelly before she can remove the sperm packet itself. It's a way of protecting the sperm.

'What's really fascinating is that a whole variety of these adaptations have evolved within crickets. In tree crickets, for example, the male produces a soupy-like secretion from specialised glands situated just under the wings. After the female has received the sperm packet, she actually feeds on these irresistible proteinaceous secretions, which takes about half an hour, during which time sperm are transferred to the female.

'There's many other sordid strategies. In some species of ground cricket, the male has a little spur on his hind legs, and he offers this up to the female's mouth during copulation. The female chews at this spur and it feeds on the male's blood. The longer she feeds on it, the longer the spermatophore is attached and the more sperm the male transfers.'

So, you've got two competing interests. The female wants to get the best sperm from the best male; the male wants his sperm to be used. All these complex behavioural interactions exist to try to achieve each of those outcomes.

'In a way there's sexual conflict – the male's reproductive interests don't necessarily align with the female's. Each individual male wants to make sure his sperm are transferred into the female, whereas the female wants to choose the best male. Sometimes these strategies that seemingly favour the male can be manipulated by the females.

'Take one of the more bizarre strategies used by the Australian Scaly Cricket, *Ornebius aperta*. They have a prolonged mating session lasting 3–4 hours, and the male engages in multiple mating. This means it doesn't matter whether the female eats the entirety of each sperm packet almost immediately after it's transferred, because the male will mate with the same female up to 58 times during this mating session.

'The amazing thing is that one experiment showed that females are able to distinguish between males in good and poor condition by engaging in this strategy, because only males that are in good condition can actually produce the full 58 spermatophores in this timeframe. Males in poorer condition were unable to mate quite so often, or at least the females gave up on them before they were able to. The female could actually use this male behavioural strategy in her cryptic female choice, deciding whether or not to stay with the male to receive the full quota of sperm packets.

'In some species of bush crickets, the females don't seem to have quite so much choice. One strategy that I've studied is prolonged copulation, where the male uses special grasping structures. These are adapted from the cerci, which are like a little pair of extra antennae and found at the end of the insect abdomen. In some species of bush crickets these have become specialised claspers. So, rather than producing a jelly-like nuptial gift for the female to eat, the male saves energy and remains attached to the female using these claspers during sperm transfer. So, by the time the male lets the female go, sperm has already been transferred into the female. If you look at bush crickets that have the specially adapted claspers, you find that the spermatophylax (the jelly-like blob) is often absent altogether. The strategy of hanging on to the female has actually replaced the sack in evolutionary terms, as a way of protecting the sperm.'

Crickets are so well known for their distinctive song: we know they have a calling song and a courtship song to encourage them. I wondered if there were any other variations.

'In field crickets the song is used in a variety of different ways. If the males engage in combat, they will use an aggressive song, which is quite loud and strident. The outcome of a fight is quite easy to judge – you can see who the winner is because he invariably jerks his body backwards and forwards like a triumphant dance and, at the same time, does a special victory song, which announces that he's the winner, and the loser then backs down. Quite often the loser will remember the identity of the one who beat them and will avoid engaging them in direct confrontation again.

This victory song is really important in establishing who is the winner of that contest.'

Crickets in the UK are not as common as they once were. But Karim believes this may be shifting, and there is hope that we might yet be able to enter a flower-rich meadow and hear a cricket calling.

'We have some very rare species, but we also have some species that are actually undergoing rapid expansion and are becoming much more common throughout the country. Crickets are warmth-loving and some opportunist species like changes in the climate. For example, Roesel's Bush Cricket and the Long-winded Conehead have been expanding in range up the country. On the other hand, some species aren't doing so well. At one point, the field cricket was limited to a population of roughly 100 individuals. Reintroduction and captive breeding have now begun to address that issue.

'If you're looking at true crickets, they're less common. Mole crickets are only found in one tiny area of the UK and may even be extinct. The scaly cricket, also known as the Atlantic Beach Cricket, is found in only three locations of the British coasts among shingle.

'When I first moved to Derbyshire 20 years ago, I couldn't find any species of bush crickets. Now I've seen and heard the Speckled Bush Cricket, the Long-winged Conehead and Roesel's Bush Cricket in my back garden. A lot of people will be able to hear them singing on sunny summer days, especially with the expansion of Roesel's Bush Cricket.'

Dying for a shag

In the insect world males can go to a lot of effort to get a girl, but sometimes it really is a fatal attraction. Males are rather expendable in the insect world, and for a male honey bee, sex and death go together. When a bee colony reaches a certain size, a number of new queens are produced. After a new queen has emerged, it takes a few days for her to become sexually mature, at which point she flies off. During this mating flight she will locate an area where male or drone bees from more than one colony have congregated – either that, or they will find her. Drones will fly after her and compete in the air to reach the flying queen and mate with her. Once a successful drone makes a mid-air coupling and ejaculates, he stops flying and lolls backwards as if unconscious – I wonder if he knows what's coming next. The queen bee flies on, carrying his weight, but the drone bee eventually falls off and drops to the ground dead – leaving behind the major part of his genitalia, still attached to the queen. The next drone has to remove this obstruction before he too can mate. The queen may mate with a dozen or more drones, and it may take more than one flight before she has stored enough sperm in her spermatheca to last her a lifetime. This is rather important, as she will lay as many as a couple of thousand eggs every day for the next 4 years.

Storing all the sperm you will ever need after a few short matings is a great example of the kind of efficiency that has made insects able to multiply so quickly. All kinds of insects seek to fine-tune the basic reproductive strategy to make it as effective as possible. For insects, it's not size that matters – it's time.

Brief encounters

Insects that do not live very long cannot really afford to take their time over mating. It's got to be done quickly, or it may never happen at all. You might have been lucky enough to witness a large mating swarm of mayflies. It can be a spectacular sight – and for the mayflies, it is the first time they will see each other, and the act of mating is the last thing they will ever do. Their brief lives as adults, which in some species can be a matter of only a few hours, means necessarily brief encounters. After feeding for a year or two, the aquatic mayfly nymphs are fully grown. Triggered by rising water temperatures, among other factors, they rise to the surface of the water and emerge as dull-coloured, winged subadults, which fly off to perch on shoreline vegetation. In a move that is unusual for insects, at this stage they moult again to become the shiny-winged adult. Once their cuticle and wings have hardened, millions of adults will take to the air to form large mating swarms.

These swarms, which have a characteristic rhythmic rising-and-falling flight pattern, take place at dawn or dusk – over water, in clearings or near obvious features in the landscape such as trees or bushes that entomologists call 'swarm markers'. Males of many species have their eyes divided into upper and lower portions, the upper portion sometimes being raised on short stalks. The enlarged facets in the upper part of the eyes help the males to see potential mates above them in a swarm. Males then use their long front legs to hang off their chosen partner, engaging their tarsi (feet) into special regions on the sides of the female's thorax. Male mayflies have paired reproductive organs which are inserted into paired vaginal openings in the female. Once successfully locked together, copulating pairs will gradually sink down out of the swarm towards the water below, where the females lay hundreds or thousands of eggs. By the morning it's all over – but what a night. What really astonishes me is that this, or something very like it, has been going on for the past 300 million years, and for most of that time there wasn't anybody around to see it.

I was in North America to film the massive simultaneous emergence of mayflies along the banks of the Mississippi in Wisconsin. Predicting exactly when a swarm will happen is a bit of a dark art, but we had taken good advice from local scientists and had even contacted the meteorological service, whose weather radar could see these swarms when they occurred. In previous years, large swarms of mayflies had even caused accidents by making the surface of the roads so slippery from their crushed bodies that drivers had lost control of their vehicles. This was going to be one of those television sequences that people remember – a piece of TV gold. Two days had passed and nothing had happened, but we were assured that we would hear as soon as a swarm was seen on the radar screens. It was just after midnight on the third day when there was a loud banging on my motel room door. I stumbled to the door and opened it to see the cameraman, Johnny Rogers, and the sound recordist, Parker Brown, hurrying down the stairs, loaded with their kit. 'We've had the call – it's already started!' they shouted as they disappeared out into the parking lot. There wasn't a second to lose. I got dressed as fast as I could and grabbed my bag and ran down the stairs two at a time. In my frantic hurry I tripped on the bottom step of the staircase and crashed headlong through the fly-screen door, out into the humid night air. Staggering to recover my balance, I looked about and could clearly see Johnny and Parker collapsed in a heap by the car, laughing so much they could barely breathe. They had not had a call – there was no mayfly emergence. 'You pair of utter bastards,' I hissed. Johnny loves playing tricks on me and this was one of his very best performances. I wasn't really cross with him; I was just bitterly disappointed that we were not going to see mayflies filling the air and piled so deep on the ground that they blocked bridges and railway tracks. We never did see a big swarm, so that spectacle remains on my 'must see before I die' list.

There's no such thing as a free meal

How many times has dinner (or even breakfast or lunch!) been a prelude to sex? You're feeling relaxed and replete and you quite like the appearance of your potential mate. It's much the same with insects. Males can give their potential mates a gift of food, known as a nuptial gift. This offering encourages the females to accept them or, at the very least, not reject them out of hand. The gift is important for the male as well, because if it is of sufficient quality, it will make sure the proceedings last long enough for him to success-fully transfer his sperm.

Dagger flies, also known as balloon flies, are common to the northern hemisphere and can be found in all sorts of habitats, from hedgerows and meadows to woodlands and wetlands. What makes them instantly recognisable is their smallish, rounded head with distinctive mouthparts in the form of a rigid, elongate, downwards-pointing proboscis. If you've never seen one you don't spend long enough looking at vegetation. I can absolutely guarantee that you will find something interesting within five or ten minutes if you simply look carefully at what's going on in a small patch of plants. You might be lucky enough to see a male dagger fly hanging by his front legs from a leaf in the act of copulating with a female. The female will be eating a small insect – a wasp, fly or planthopper – while she mates. So, what led to this happy rendezvous? Before they hitched up, the male had gone hunting for a nuptial gift – something tasty to offer any potential mate. He might have met another male dagger fly that had already secured an item of prey and, rather than go to the trouble of finding his own, a tussle might have ensued. At any rate, the male is now ready to go a-wooing. His chances of success would be dismal if he turned up empty-handed. The size and quality of the prey shows the female that she might have met a suitable mate. With preliminaries over, and while the female is tucking into her bug tartare, the male gets down to business.

A problem for many male insects is keeping the female interested and occupied for long enough that he can successfully transfer his sperm. So to add a little something extra, the males of some dagger fly species wrap their nuptial gifts in silk produced by special cells in the tarsi of their front legs. These balloons of silk-wrapped snacks can be a wildly successful strategy, for as the female probes and unpacks the gift on which she will dine, the males have bought themselves extra time to mate. When mating is completed, the male may take what's left of the gift away from the female and try to use it again with a new mate. In fact, he may use it a few times until he has to throw it away and find a more appetising treat. As a character in the 1954 radio drama *Under Milk Wood* by Dylan Thomas observes, 'Men are brutes on the quiet.'

But then, consider the subterfuge of a North American species who, rather than go to all the bother of catching anything, merely offers his mate a large silk balloon with nothing in it at all. Well, that's not quite fair, it's not really like an empty gift-wrapped box – internally, it's quite a complex structure and is probably a good demonstration of the male's quality as a suitor, even if the female does not get a free meal.

Nuptial gifts of prey items are one thing, but what if the nuptial gift is actually a part of yourself?

Sexual cannibalism

Some North American sagebrush crickets have taken this approach to a whole new level – they actually get bits of themselves chewed off. These large, chunky crickets live in the subalpine sagebrush meadows and pines of Wisconsin and Colorado. Courtship begins in typical cricket style, with a song the male makes by rubbing his toughened front wings over each other. But once a female has shown interest, something rather unusual happens.

She mounts the back of the male and tries to eat the fleshy, lobe-like hind wings of her mate. She chews off chunks and laps up the body fluids that leak from his wounds. For the female, these protein-rich love bites are worth having, as they will help her mature her eggs. For the male, it's not so bad either – and rather than fight her off, the male submits to her cannibalism. He even holds up his front wings to make it easier for her to feast, but not before he has made sure he has coupled his genitals with hers. To keep her firmly in place, he grabs the end of her abdomen tightly with a special hooked clamp and will only release her when his sperm is safely transferred. There are understandable reasons why he needs to hang on tightly to the female. The cost of sex to the male is high. When they mate, male sagebrush crickets may lose up to 10% of their weight, so they don't want any freeloading female getting a meal without having sex. The males have to get it right because once they've mated, and are partially chewed, they are 'damaged goods' and are not going to be nearly as attractive to other females.

It does always seem that when an insect is doing anything out of the ordinary, it will be to do with sex, the result of which is to leave behind as many viable offspring as possible. And what feature is more unusual than the ability to glow in the dark?

Seeing the light

Growing up in Scotland, I never saw a Common Glow-worm, *Lampyris noctiluca*. Although much less abundant the further north you go, this beetle is found from the west coast of Ireland right across Europe and Asia, to as far as China. It was only when I moved south that I saw glow-worms from time to time. They are recorded in field guides as a common species, but like many other insects, they are much less common today than they once were. The males and females look markedly different. The males are normal-looking, brownish, elongate beetles. The wingless females, which are larger than the males, are often described as looking like miniature trilobites. Glow-worms live in a variety of well-vegetated habitats, where they feed on snails and slugs. Most of the year you would never know they were there without grubbing around hunting for them, but it is during the summer months when you can see them at their best. At night the females climb up a plant stem and start to glow from two patches of transparent cuticle on the underside of their abdomens. Their aim is to attract males to come and mate with them.

The female glow-worm waits until it is dark and then curls her abdomen round to expose her underside, and may also wave it gently to and fro. The eyes of male glow-worms are large and highly sensitive to the light the female emits. Once they mate the female stops glowing and will then lay up to a hundred eggs in the soil or among leaf litter over the course of a few days.

Fireflies are closely related to glow-worms, and watching them twinkling as they flutter through the darkness is surely one of the most entrancing sights in the natural world. I've been lucky enough to see them in several parts of the world and once even collected enough in a glass jar that I could read a book by their light. They have large eyes, but their head is usually hidden by a hood-like extension of the thorax. On the underside of the abdomen, underneath patches of transparent cuticle, they have special luminous organs that, just like glow-worms, produce a cold, greenish light. Rather than simply glowing, these beetles are able to control the interval and duration of the flashes they make by regulating the amount of oxygen supplied to the luminous organs.

In the normal course of events, a male will fly around flashing unambiguous, species-specific bursts of light in the hope that a female firefly perched on low vegetation will flash back at him. Depending on how far the male is from the female, this reciprocal light show will be repeated until the female firefly guides her suitor ever closer, whereupon they meet and mate. But this seemingly foolproof approach can be hijacked and, for some unlucky fireflies, sex can be an incredibly risky business. In North America there are some large fireflies that prey on smaller fireflies. If a female of the larger firefly feels like having sex, she will respond to the flashes of specific males, giving the correct signal. However, if she is hungry, she will imitate the flashes of a female of the smaller firefly to lure the males to their death. Because of this cunning strategy, they are known as the femmes fatales of the insect world.

Don't let the bed bugs bite

A few years ago on a filming trip to Christmas Island, I was woken in the middle of the night and could sense something moving about slowly on my chest and neck. I have learned from experience never to go anywhere without a head-torch and, a little later, I discovered that several bed bugs had been feeding on me. One of them, its abdomen now completely swollen with my blood, was moving slowly off to find a hiding place. I quickly grabbed it and, by turning the bed frame over, I gathered a few more of the apple-pip-sized creatures into a small collecting tube. I presented them to the hotel manager the next morning. 'Aw yeah, mate, and how d'you know these are actually bed bugs?' he drawled, without showing any interest. I showed him my Oxford University staff card and immediately, his demeanour changed. He seemed more concerned that I might mention it to the local tourist office. I didn't tell the tourist office, and that evening the entire film crew enjoyed several bottles of Australia's finest Cabernet Sauvignon. Pretty fair considering the relative volumes of red liquid exchanged.

Adult bed bugs are flattened, reddish brown, wingless, nocturnally active insects, but what makes them really interesting is, yes, their sexual behaviour. Rather than having a 'standard' insect penis, the functional part of a male bed bug's genitalia is like a stiletto dagger. Stabbing your partner and injecting sperm is known as hypodermic or traumatic insemination. This technique means that the males, freed from the bother of all the courtship malarkey, simply go around stabbing females willy-nilly and they don't even have to worry too much about where they stab, either. The sperm migrate through the female's bodily fluids to reach the ovaries.

Female bed bugs can be damaged by this kind of insemination and can suffer immune reactions or bacterial infections. However, evolution has provided the females with some degree of protection in the form of a special organ located on the underside of the abdomen, whose function is to receive sperm directly and reduce the impact of repeated inseminations. This special organ consists of a groove on the hind margin of the fifth abdominal segment, with a sac-like structure beneath. To make it easy for the males, it is clearly visible from the outside and, just like the 'tear here' instructions on packaging, it encourages males to use the 'easy access' groove for penetration rather than randomly peppering the females with holes.

New men

The occurrence of female brood-guarding is seen in many insects, but is characteristic of treehoppers. Females guard their eggs or young nymphs with vigorous wing-fanning, buzzing, rapid movements and even physical contact. Female treehoppers have to defend their eggs from things that would simply eat them, but also from some much sneakier enemies that try to fly in and lay their own eggs inside the treehopper's eggs. It can be full-time job, but the effort is definitely worth it: a much higher percentage of the eggs will survive and hatch than if they were left unattended. These days we expect male humans to take on their fair share of childcare, but humans have got a long way to go before they can match the behaviour of some insects.

The most famous case of sexual role reversal is to be found in giant water bugs. Here, unusually in the insect kingdom, it is the females who take on the job of courtship. They compete for available males, on to whose folded front wings they glue their eggs. If there are few males without eggs around, things can turn nasty. Females will approach males and try to swipe the eggs off his wings. The male will fight back, but if the female succeeds, he then meekly accepts defeat and mates with her and the victorious female will then lay her own eggs on his wings. The cost to the male of this brooding service is quite high. While carrying eggs he cannot fly and is less able to catch prey. Because of this high cost, it is important that the male is sure that the eggs he is carrying were fertilised by him and not some other male. And so a male giant water bug will only allow the female to glue her eggs to his wings once she has had sex with him and even then, he will only allow the female to glue a few eggs at a time before he copulates with her again. When all the eggs are laid, the female leaves the male on his own to look after their brood.

Of all the superpowers insects have, their phenomenal powers of reproduction are vital. The faster they breed, the quicker they can adapt to environmental changes, and the more they successfully reproduce, the better are their offspring's chances of survival.

The sexual antics of insects are truly fascinating, and their life cycles are even more interesting. Many insects chew green plants. Many other insects chew on the insects that chew on green plants. In the next chapter, we'll find out about a large group of insects whose young eat other creatures from the inside out, and why a certain type of wasp made Charles Darwin question his belief in the existence of a caring and compassionate God.

Chapter 5

THE BODY-SNATCHERS

Flesh-eaters

It was October 2007 and we had just finished filming *The Lost Land of the Jaguar*, the second of a BBC expedition series. It had been a gruelling six weeks in the humid interior of Guyana and the members of the team were tired out and preparing to head back to the UK. My friend Johnny Rogers, the camera operator, was complaining about a boil just above his left ear. He had had it for about four weeks. It had started off small – like a pimple or a whitehead, but now it was causing him quite a lot of discomfort. It was keeping him awake at night and he could feel occasional sharp pains and even hear scratching noises. When he said he could feel something moving I knew right away that he had the larva of a bot fly inside his skin. A close examination of the boil with my hand lens revealed the unmistakable rear end of a maggot-like creature. Two little spots, the breathing holes of the larval tracheal system, were poking out. 'Oh, this is great!' I exclaimed, 'It's a bot fly larva.'

Now, it's a fact that, when working in the wilds of South America, most entomologists worth their salt claim to want to be host to a bot fly so that they can enjoy its development first-hand, and I have to confess that I felt rather jealous. Johnny was not at all impressed by my excitement and demanded that the 'little fecker' be removed post-haste. One of the crew, who had rather long fingernails and some experience of bot fly removal, got to work and a little while later, a bot fly larva wriggled its last in a small pool of Johnny's blood. The larva was pale, with a pair of sharply hooked mandibles on the head and bands of short, stiff spines circling the segments of the body, which served to anchor it in place and prevent it from being dislodged. One bot fly eating your flesh is bad enough, but imagine if you were a wild animal with 10 or 50 or as many as a hundred larvae eating you. Frankly, it doesn't bear thinking about.

The bot fly that was in my friend's head is widespread in Central America and the warmer parts of South America. The larvae of the Human Bot Fly, as it is known, develop in the skin of a number of wild and domesticated animals. Happily, it is the only one of its kind that parasitises humans. These bot flies are a little larger than a house fly, and you might wonder just how they could lay an egg on your head or on your face without being detected. But it is not the female bot fly that makes the choice of where her eggs are laid, and this is where it gets really interesting. They have evolved a very clever strategy. The females catch blood-feeding flies such as mosquitoes in mid-air. More than 40 blood-feeding species are used unwittingly as bot fly 'egg-carriers'. Once the female bot fly catches a suitable mosquito, she holds it gently and lays a small clutch of eggs, usually on the abdomen.

The surprised mosquito is then released unharmed – and for the human bot fly, that's as far as parenting goes. When the mosquito next feeds on a suitable host, the warmth and odour of the skin causes a bot fly egg to hatch, and a tiny larva burrows into its food source as fast as possible, most likely using the puncture mark left by the blood-feeding mosquito. Here, it will chew and rasp away, getting bigger and bigger until one day, eight weeks or so afterwards, it will pop out, drop to the ground and pupate. Myiasis is the technical term for the condition where the larvae of insects burrow into animals (and humans).

Before one trip to East Africa, I was told by those more experienced to beware of this or that creature. Nevertheless, no one told me about the Tumbu Fly, a species of blow fly common in subtropical Africa. A good way to avoid having a Tumbu Fly larva burrowing in your skin is to always iron your clothes. This not just a story made up by old colonial army types to ensure 'standards in the field', but is really because the fly concerned – a rather nondescript blow fly – likes to lays its eggs on damp laundry. If the eggs are not killed by ironing, they will hatch on contact with the skin, especially in areas where the cloth is in close bodily contact, such as waistbands and hatbands. I saw a hysterical headline a few years back which screamed, 'Maggots Ate My Face!' The truth was slightly less scary. A golfer living in Hull was returning from an African trip and had worn a cap that had clearly been visited previously by a female Tumbu Fly. An egg had hatched while he wore the cap and by the time he returned to the UK, he had a nice big boil in the middle of his forehead.

But I feel really sorry for the common toad, which has even more to put up with. A very specialised blow fly called the Toad Fly lays its eggs on the back and sides of the poor old toad, and the hatched larvae wriggle their way towards the toad's nostrils. Here, they take up residence and start to gnaw away at the toad's flesh. As the maggots grow they eat more and more tissue and, depending on their number, can even consume most of the toad's head. Death is inevitable. In some parts of the northern hemisphere, where these flies are found, well over half of all toads can be parasitised in this gruesome manner. Another interesting effect, seen in many other parasitised animals, is that the affected toads change their behaviour and, rather than keeping themselves to nice, damp, sheltered spots, they move out into the open, where even more toad flies will find them an irresistible place to lay their eggs. When fully grown, the toad fly larvae will leave the corpse and wriggle away to pupate in the soil.

Slow death

Makers of natural history programmes used to leave out nature's 'nasty bits'. Audiences did not generally like seeing furry animals getting ripped asunder. But things are getting a little more realistic these days. Viewers might get to see a doe-eyed deer polished off by a predator and the once all-conquering Monarch of the Glen staggering off to die alone, mortally wounded by a younger and stronger rival. By and large, audiences are fine with the basic concept of nature 'red in tooth and claw', just as long as they don't have to watch it. Many of us, it seems, prefer our experience of nature to be censored and sanitised. But the natural world is not kind or considerate. It's not mean or malign, either. If nature is anything at all, it is simply indifferent. And nothing illustrates this more than the world of parasites. If you are easily grossed out, perhaps you might want to skip the rest of this chapter.

There are few worse ways to die than being eaten alive. But one of them is being eaten alive very slowly. However, this is the precise fate of a large number of insects. If they are not directly gobbled up by birds and reptiles or one of a whole host of mammalian insectivores, whose populations they support, they may suffer a protracted death at the hands, or rather the mandibles, of other insects – especially parasitic wasps and flies. Before I go any further, I must make it absolutely clear that when I say 'wasp', I do not mean the striped, loudly buzzing creatures that build paper-carton nests and are attracted to fallen apples or the jam on cream scones. The overwhelming majority of wasps belong to many different families and are a different kettle of fish altogether – they are parasitoids. That is, they lay their eggs on or in the tissues of a host animal, typically an insect or spider. The wasp eggs hatch and the larvae eat the host alive. The word 'parasitoid' is used to distinguish this lifestyle from that of a parasite, where killing the host is not really such a good idea. In the UK there are nearly 8,000 species of wasps and the majority of these (almost 6,000 species) are parasitoids.

It was the life cycle of wasps like this that caused Charles Darwin a great deal of anxiety. A year after the publication of *On the Origin of Species* (1859), Darwin wrote to his friend and supporter, the American naturalist and professor of botany at Harvard University, Asa Gray. Among other things, he wanted to know if there had been any good reviews of his book. Growing up, Darwin had considered all the careers open to a man of his position in society – law, medicine and the church – but he much preferred to go 'entomologising', causing his father to despair of him ever doing anything useful. His extensive knowledge of the natural world and his experiences on board HMS *Beagle* had brought him to the point where he understood how species arose, and it flew in the face of the accepted biblical version of events. In the letter to Asa Gray, Darwin reveals something of his mental turmoil in this now famous sentence:

> *I cannot persuade myself that a beneficent and omnipotent God would have designedly created the Ichneumonidae with the express intention of their feeding within the living bodies of caterpillars.*

Darwin further confesses that the whole subject is too profound for the human intellect and adds, 'A dog might as well speculate on the mind of Newton.' He writes that he finds himself bewildered, and protests to Gray that his views are not at all necessarily atheistical. Nevertheless, there can be little doubt that his first-hand knowledge of the life cycles of ichneumon wasps had shaken Darwin's belief in the existence of a creator considerably and, in the end, he simply suggests that every man should hope and believe what he can.

So, what of the Ichneumonidae, the large family of some 25,000 described species, that brought about Darwin's near apostasy? Incidentally, many experts believe that number of known species is only about a quarter of the species that actually exist. These wasps, and many others like them that consume and kill their hosts, are a vital part of the natural processes that regulate insect populations. Without this control, the numbers of certain species would increase greatly – overwhelming host plants and upsetting the balance. It is not planned, of course, but stability seems to be more favoured over boom and bust. Wherever there is a living to be made, it will be made.

Wasp waists

Have you ever wondered why wasps have an incredibly, sometimes unfeasibly, narrow waist? For evolution to force the nerves, the gut and other structures through this restricted passage might seem odd at first. The answer lies at the end of the abdomen, in the form of an egg-laying structure called the ovipositor. The narrow waist makes the abdomen of wasps highly manoeuvrable and capable of getting to just about anywhere it can reach. I forgot this crucial fact when trying to hold a large ichneumon wasp gently by the wings for a close examination. She deftly swung her abdomen round and rotated it so that she could plunge her ovipositor into my finger. It must have found a nerve because a sharp stab of pain shot through my finger and into my hand. Interestingly, the males have the same reaction but, not having the right equipment, it's little more than hopeful posturing.

The ovipositor of a parasitic wasp is a lot more than a simple hollow hypodermic. It is an extremely complex bit of evolutionary engineering and is made up of two pairs of slender elements called valves, which often have serrated tips. The valves are joined together along their entire length by a tongue-and-groove mechanism similar to one you might find on a ziplock plastic bag, and which can slide against each other as they are inserted into the host animal. The hollow space in the middle of the valves is where the wasp's eggs move down the length of the ovipositor. The stinging apparatus of social wasps and bees is an ovipositor that has evolved a different role, that of injecting venom into enemies.

Living larders

Everyone will be familiar with the term 'parasite', meaning a species that lives at the expense of another but does not cause its death. Blood-sucking fleas and lice, as well as the human bot fly, are good examples. But parasitoids are different. For these species, the host is nothing more than a quantity of live tissue and energy on which their young can develop – but they must be careful to select a host that is large enough. There would be no point in laying your egg in a host that was too small for the young to complete their development. Remember the scene in the film *Alien* (1979) where Executive Officer Thomas Kane, played by John

Hurt, suddenly clutches his chest? Almost certainly based on the life cycles of real parasitoid wasps, the alien creature that bursts from his rib cage was an endoparasitoid – that is to say, it ate its host from the inside and emerged when fully grown. In reality, the character played by Hurt would have been feeling pretty unwell for quite a while before his grisly end. The opposite of this strategy is to be an ecto-parasitoid and cling on to the outside of your host while eating it – it's a riskier undertaking, unless the host is immobilised. Had the alien of *Alien* been an ectoparasitoid, Executive Officer Kane would have been only too aware of it as it fed, perhaps wrapped round his neck, and would have disposed of it in good time. In the real world, just as in science fiction, the two strategies need a different approach. Ectoparasitoid wasps cannot risk their young being knocked about or groomed off, so they paralyse the host and hide it in a burrow or somewhere concealed before laying their eggs. The host simply becomes a living larder – all the wasp has to do is to make sure there is enough food in the first place.

Endoparasitoids tend to allow the host to live on as normal and even grow after its eggs have been laid. This is where things get really clever. Let's imagine that the host is a caterpillar. The caterpillar needs all its organ systems to function; so there would be no point if an internal wasp larva suddenly chomped its way through the gut or some other major organ or nerve. The caterpillar would die, and decay, and the wasp larva would not be able to complete its devel-opment. We still do not fully understand the exact nature of what happens inside a host, but as the parasitoid larva grows, it avoids eating anything that will seriously damage the caterpillar's ability to feed and grow. It also means that the host can behave normally and avoid being eaten by other things. Meanwhile, the wasp larva quietly feeds and grows inside, perhaps eating just fat tissue and fluids. But when it reaches a certain size and 'knows' the finishing line is in sight, it goes on a feeding frenzy. Whatever mechanism constrained its feeding before is now switched off and the larva will quickly devour the caterpillar from end to end before it emerges to pupate, often attached to the shrivelled corpse of its host. But what if another parasitoid wasp had come along a few hours after the first wasp had laid her egg and then laid an egg of her own? If this happened, the egg or hatched larva of the second wasp would be at risk of being eaten by the now larger larva of the first wasp. To avoid this, the second wasp may be able to pick up the odour left behind by the first wasp so that she finds another victim.

Carnivorous and herbivorous animals are rather dull. They simply encounter their food and eat it. It's all pretty straightforward. But the life cycles of parasites are incredibly diverse, and some are exquisitely complicated. The parasitic lifestyle is very common indeed; you might be excused for thinking that this sort of thing only happens in sultry jungle interiors, where nature runs amok, but it happens everywhere. The countryside around you may appear nice and peaceful, a tranquil idyll, but it is a biological battleground where just about anything goes and nobody is safe. Even the aquatic pupae of some caddisflies can be parasitised by a tiny wasp that has evolved the ability to swim down below the surface to lay her eggs inside their protective cases.

A bug's life

One of the most familiar white butterflies is the Large White. It can be serious pest of a large range of field crops such as cabbage, broccoli and cauliflower. But lurking in the undergrowth is a small parasitoid wasp belonging to a large family called the Braconidae, second only in numbers of species to the Ichneumonidae.

It all starts well enough for the Large White butterfly. The slightly sulphurous smell of cabbage-like plants attracts the attention of a female, and she lays a cluster of eggs on a leaf. The eggs hatch and the young caterpillars start to feed – but by feeding, they may well be sealing their own fate. Many plants give off airborne chemicals in response to being munched, and these chemicals are known to attract parasitoid wasps. It is as if the plants are calling in a SWAT team. But the plants can even enlist the help of specialist troops, as they are able to subtly alter the ratios of different compounds in the chemical mix in response to different sorts of herbivores, whether they be leaf-chewers such as white butterfly caterpillars, sap-suckers like aphids, or generalist grazers like slugs.

Alerted by the faint whiff that signals the presence of chewing caterpillars, a female braconid wasp wastes no time in locating the source of the odour. Using her ovipositor, the female wasp lays eggs in the caterpillars she finds and, depending on their size, she may inject a few eggs or as many as 50 in each. As the wasp lays her eggs, the caterpillar reacts by thrashing its head from side to side but, once the ordeal is over, it will resume feeding. In about two or three weeks, the parasitised caterpillar will wander off to die and, not long afterwards, the fully grown wasp larvae will chew their way out through the skin of the caterpillar, and each will spin a silken cocoon inside which they will pupate.

But the game is not over yet, for there is an even smaller, rarer wasp that wants a piece of the action. This one is a parasitoid of parasitoids, and the females, using airborne chemicals as a homing beacon, seek out the cocoons of the first wasp, in which they lay their own eggs. If you ever find the remains of a parasitised butterfly caterpillar with wasp cocoons attached, you might like to pop it into a jam jar covered with muslin and see what emerges. No one knows for sure exactly how may parasitoids there are in the world, but it may well be that a full quarter of 'creation' exists to consume the rest of 'creation' alive. No wonder Darwin had his doubts.

No hiding place

Sawflies, which are not flies, are the vegetarian close cousins of wasps. As they lay their eggs in plant tissue – which, by and large, stays put and does not fight back – they have no need of a narrow waist. The largest sawfly in the UK is the Giant Horntail or woodwasp. The females are impressive beasts that can reach 4 centimetres long. They have distinctive black and yellow banding and are sometimes mistaken for a hornet. The large ovipositor and sharp spine-like projection at the end of the abdomen, the horn from which it gets its common name, are more than enough to convince many birds and most people that this is an insect to be feared. While on holiday one year in Kintyre on the west coast of Scotland, I remember my father picking up his artist's stool and flailing it about madly at an approaching woodwasp. I tried in vain to tell him it was not going to sting him or me, but nonetheless he kept on swiping at it until it had disappeared.

The horntail is widespread across the UK wherever pine and other softwood trees are grown and felled. I got a wonderful opportunity to spend time with horntails when making a short film about them. Female horntails like to lay their eggs in fallen or weakened trees as well as in cut logs. As they lay each egg using their stout, saw-like ovipositor, they introduce the spores of a particular fungus they carry in a pair of special sacs located inside the abdomen. The female horntail also injects a special slime to encourage the growth of the fungus as it spreads into the wood. A couple of weeks after being laid, the horntail larvae hatch out and will spend up to three years burrowing through and eating the fungus-infected wood before they move to the surface to pupate just below the bark.

Of course, a succulent food source like a horntail larva, even encased in wood, will not be completely safe – there is a parasitic wasp that targets them. The graceful, dark-coloured sabre wasp is the largest ichneumon wasp in Europe. The females can reach an impressive 40 millimetres in length, but what makes them really remarkable is the length of their slender ovipositors, which can be as long again. The female sabre wasp uses her antennae to detect faint odours coming from the burrowing woodwasp larvae and, once satisfied that a particular spot is worth her while, she starts to probe the surface of the wood with her ovipositor. She begins to drill, and as she does, the protective sheath of her ovipositor folds back, revealing just how slender the ovipositor really is. You really have to wonder how she is able to get such a fragile-looking structure through solid wood.

Standing almost on 'tiptoes' and with her abdomen held high, the wasp begins to penetrate the surface of the wood. If her ovipositor was too stiff and brittle, it would break easily, like a strand of uncooked spaghetti. If it was not stiff enough, it would be like trying to push a bit of cooked spaghetti into a slot machine. But her ovipositor has evolved to be just stiff enough and just flexible enough to do the job – most of the time. The cuticle of the ovipositor is also hardened by the addition of metal ions such as zinc and manganese. Drilling is quite a slow process, and the wasp might not get the tip of her ovipositor in just the right place the first time. She may have to withdraw and try several times before she gets close to her prize – a juicy horntail larva to paralyse and lay her egg on. The larvae of sabre wasps are ectoparasitoids, and the horntail larva is gradually eaten alive from the outside. The time it takes her to drill and lay her eggs puts the sabre wasp herself in danger, and I did see one or two ovipositors sticking out of the wood. The wasps they belonged to were eaten by a bird, and the energy their bodies contained became part of the next level of the ecological pyramid.

No matter how big or well defended you are, you can never depend on being safe all of the time. There is the constant danger that something else is going to eat you or breed inside you.

The stinger stung

For even the most fearsome insects in the world, a parasitoid may be their undoing. In my profession it would not do to be that worried about being stung by insects, but there is one species I would go out of my way to avoid. I've often watched them running up and down the bark of trees in South American rainforests, and I've wondered what the pain would really be like if I let one of them sting me. Once, nearly overcome by scientific curiosity, I even reached out to grab one, but thought better of it. The species in question is a large hunting ant known as the Bullet Ant. Their sting, which is considered to be the most painful sting of all, has been likened to having a red-hot nail hammered into your flesh or being shot – hence their common name.

There are certain South American tribes who use bullet ants as a test of manhood. This dangerous ritual involves collecting dozens of these ants and securing them in a mat or glove made out of woven plant fibres. Teenage boys wishing to be considered men thrust their hands into an ant glove or have a mat of ants slapped on to their chest. They must endure this ordeal not once, but as many as 20 times over the course of many months. The only person I know who has willingly suffered this torture is the British naturalist and broadcaster Steve Backshall. When I asked him about the experience, he described how, at first, his heart rate accelerated and then he started to sweat. Within 15 minutes there was nothing but overwhelming pain. He began to slip in and out of consciousness and suffered hallucinations. The locals encouraged him to let go and not to fight the pain. Crying made him feel a little better. Even three hours after the ceremony, he said it felt like someone was pounding his hands with hammers. Interestingly, and probably as a result of the endorphins produced in response to the pain, he said he felt superhuman – though I always knew he was.

The active ingredient of the sting, known as poneratoxin, is a powerful neurotoxin that paralyses muscles and generates prolonged pain signals, which are relayed to the brain. Would I have subjected myself to this agonising torment had I been asked do it for a documentary? Umm … probably. The things people, including myself, do for television never ceases to amaze me. I did have a minor altercation with another South American ant species – the army ant. These ants are swarm-raiders that scour the forest floor looking for food, and they form temporary nests called bivouacs. We had found a bivouac in a hollow tree trunk and I thought it would be a good idea to get a probe camera down inside to get a better look at where they had set up camp.

The soldiers of this species are large and formidable-looking, with huge heads bearing enormous, curved mandibles. As they swarmed up the cable of the camera I had pushed into the centre of the bivouac, I realised the fatal flaw in my plan. The camera cable connected them to me in a direct line – an ant expressway to my flesh. How could I have been so dim? In just a few seconds, my hands and arms were covered with a hundred or more furious solider ants. They drove their jaws in and would not let go. I fled the scene pretty quickly and spent the next half hour picking them from my skin and clothes.

Tough creatures like bullet ants and army ants do have an adversary from which there is no escape. And it's not some ironclad beast or a sharp-eyed and equally sharp-beaked bird, but a really tiny fly.

Tiny assassins

Scuttle flies, named for their rapid running action, are a family of small humpbacked flies that look a bit like the fruit flies used in laboratories all over the world. These flies have highly diverse life cycles and their larvae can be found in just about every conceivable habitat. Many feed on decaying organic material, while others are predators or parasites. Quite a few scuttle flies are parasitoids of ants, and many kill their hosts in a rather interesting manner – they decapitate them. When I first learned about the habits of these little flies, I imagined them carrying tiny hatchets with which to do the deed. That's not how it works, of course. In the case of those attacking bullet ants, the female scuttle flies land on the ant's back and lay their eggs through the soft membrane between two abdominal segments. The ants are aware of what is happening and will try to dislodge the flies. In some cases, the flies will preferentially target injured bullet ants, but can also attack healthy individuals. Bullet ants can reach up to 3 centimetres in length, whereas the flies that attack them are no more than 2 millimetres long. It's a real-life David and Goliath story in miniature. The hatched larvae move forward and end up feasting on the insides of the ants' heads, which eventually fall off. How are the mighty fallen!

Ants all over the world get nailed by scuttle flies and, in most cases, the flies are specific to one ant species. The approach used by one kind of fly, which attacks the infamous and invasive fire ant, illustrates the technique they use. The female fly hovers over the foraging workers before darting in to inject an egg into the thorax of an ant. After it hatches, the larva will move into the head, where it will initially feed on fluids. As the larva grows it will eventually consume all the juicy jaw muscles and the brain, and this causes the ant to wander about in an understandably purposeless manner for a few days. The larva then pushes off the ant's jaws and the front part of its own body becomes toughened to block the hole where the mouthparts of the ant once were. Enzymes then dissolve the membranes attaching the head to the thorax, and the head that encloses the safely pupating fly simply drops off. When ready to emerge, the fly pushes its way out of the lifeless head capsule which has protected it.

Foraging worker leaf-cutter ants are at great risk of being attacked by these parasitic flies, but they are not totally defenceless. To get to the workers, the flies need to land on the leaf fragments the worker ants are carrying back to the nest. But members of a special caste of workers, called minims, are much smaller than the major workers. They sit on top of the leaf loads, keep watch for flies above and dislodge any that do land. The extra weight of the minims 'riding shotgun' does not add that much to the workload and, in any case, a little more work is surely a small price to pay to avoid losing your head.

Parasitoids can find their hosts in a huge number of ways. They may be attracted to the smell of the food that their host likes to eat or they may seek out the sort of habitat that they are likely to be found in. Once in the right area, they may use odours produced by the host insect themselves. The defensive secretions made by stink bugs might save them from a few larger enemies, but may also act as a lure for parasitic flies. Insect excrement, known as frass, is often used by parasitoids as a cue – the insect that produced it might be close by. The parasitoids might not have to travel too far, but if their host is highly mobile or disperses to new locations, they might have a slightly more difficult task. It's a game of hide-and-seek in which the stakes are high for both parties.

Bees besieged

We are slowly coming to realise just how important bees are to our food supply, and indeed to our survival, so it's high time we appreciated that these indispensable insects have a lot to cope with. Hit hard by vast amounts of pesticides and the loss of flower-rich areas, bees also face a great number of natural enemies. Attacked by anything from viruses, bacteria and fungi to spiders, hornets and birds, they really don't have an easy life. But the specialist enemies that target bees are among the most interesting.

Oil beetles are a large, dark-coloured and flightless species with a truly fascinating life history. The Black Oil Beetle, *Meloe proscrabeus*, is widespread right across Europe and Asia, although it is more common where the climate is milder. Mating takes place on warm days in early spring. The male uses its curiously bent antennae to embrace and keep hold of the much larger female, whose swollen abdomen shows that she is full of eggs. After mating, the female will lay many hundreds or thousands of eggs in a small vertical burrow she carefully digs in loose soil. The hatching of her eggs is timed to coincide with the appearance of certain solitary bees.

The hatched larvae, called triungulins, that emerge from the burrow are very active, and have three pairs of tiny legs with which they climb up the nearest flower stem. Dozens of them crowd on a single flower. Here, they sit and wait for the arrival of a bee. As soon as a bee arrives, they scurry as fast as their little legs can carry them and climb onto the back of the bee. Many will be knocked off, but enough will find a safe position where it is difficult for the bee to groom them off, and they will hang on for what will be quite literally the ride of their lives. The solitary bee to which they are clinging has been foraging for pollen and nectar to stock the underground cells in which her own young will develop. But, of course, the bee is now flying back to her nest with a number of oil beetle larvae aboard and, as soon as she is inside, the triungulins will disembark and make their way into the cells that contain the bee larvae. Here, they will devour the provisions stored by the bee as well as her eggs and young. The fully grown oil beetle larvae will stay safe underground during the winter months and pupate early the following year. All of this behaviour is, to a large degree, programmed and, although it goes wrong sometimes – for instance, if the triungulins jump onto an unsuitable host insect – it goes right often enough to work. Even after a lifetime watching insects, I still marvel at the elegance and complexity of their lives.

But oil beetles are not the only sneaky enemies bees have to deal with. Flies known commonly as thick-headed flies are mostly endoparasitoids of bees and social wasps, although some species are known to attack crickets, cockroaches and other flies. The adults are often bee- or wasp-like, but this is for protection from their own enemies and not to fool their hosts. They are often found feeding on nectar at flowers and, as bees and wasps also feed here, the flies are well positioned to find a host insect in which to lay their eggs. The technique of one species known as the Waisted Bee-grabber, *Physo-cephala rufipes*, which parasitises a number of bumble bee species, is characteristic of these flies. The female fly buzzes around and, when it intercepts a bee in flight, grabs it firmly. During the brief altercation, they may fall to the ground. The abdomen of the fly is curved and specially shaped to lay eggs inside the abdomen of the bees. Some say that the structures on the abdomen of the fly work like a can opener, but I think this is misleading. The action is more like a surgical retractor or wound spreader, and works by locking on to the edges of two of the bee's abdominal segments to prise them apart. This action makes inserting an egg through the interseg-mental membrane a piece of cake.

It's all over in a flash and, once the egg is safely inside, the fly releases the bee to go about its business. But things are not going to go well for the bee. The fly larvae will start by eating the bee's bodily fluids and fat tissues before going on to eat her ovaries. The bee is alive even up to the point when the entire contents of its abdomen have been consumed.

The life cycles of these fascinating flies are not fully understood, but like many parasite–host interactions, the larvae are able to manipulate the behaviour of their hosts. In one case, a parasitoid of the Buff-tailed Bumble Bee, *Bombus terrestris,* causes its host to burrow into the soil just before it dies. Inside the buried bumble bee, the fly pupates inside a puparium, which is in fact the hardened skin of the last larval stage. This stealthy bit of mind control is highly advantageous to the fly, because it now has a safe place in which it can spend the cold months of winter. The flies that hatch out the next spring from buried bees are bigger and have fewer wing malformations than those that spent the winter lying around on the surface of the ground.

The bumble bees can fight back a bit, and it has been shown that bees that have been parasitised tend not to return to their underground nest but stay outside in the cold all night. This is believed to slow the growth of the fly larva inside the bee and may increase the bee's chances of survival by decreasing the chances of the fly developing successfully.

Evolution by natural selection is constantly refining these complex interactions based on one thing and one thing only – survival.

Into the unknown

We really know so little of the fine details of insects' lives, and the discovery of what really goes in this miniature world has been a source of constant fascination to me. I have only touched on a few of the many examples of insect life cycles that some people may find unpleasant or even gruesome, and there are plenty more. But remember that nature is not cold and unfeeling. It is merely functional – and parasites and parasitoids are a part of the ecological machinery that keeps the world ticking over. We have learned to use them in the biological control of many other insects that would other- wise feast on the plants we grow to feed ourselves. Some of the smallest parasitoids in the world, tiny wasps mostly less than 1 millimetre in length, lay their eggs directly inside the eggs of butterflies, beetles, bugs, flies, and other wasps. Their larval development can take as little as three days, and, by killing eggs, the plants that they were laid on suffer no damage whatsoever. You'd be lucky to see one of these improbably small wasps in the wild, but if you care to look, you will find them. You could collect some insect eggs and keep them in a little pot covered with fine muslin and you might be surprised by what hatches out.

Children have asked me if I was an insect what sort of insect I'd like to be. An antlion larva is a ferocious-looking beast with huge spiked jaws that lives at the bottom of a pit it digs in sandy soil. Here, it sits and waits for something tasty to lose its footing and tumble in. The antlion larva even flicks salvos of sand grains to knock its next meal into the pit of doom. A pretty safe way of life, you might think, but there is a wasp that hunts specifically for antlion larvae. It has strong hind legs that it uses to brace the antlion's gin-trap-like jaws apart while it lays an egg through the membranes just behind its head. With the plethora of parasites that exist specifically to gorge themselves on live insects, I'm not sure I'd like to be any insect at all.

Life and death are opposite sides of the same coin. You can't have one without the other. In the next chapter we will enter a world of death and decay to discover the secret of life itself.

Chapter 6

THE AFTERLIFE

The great recycling plant

I have always enjoyed walking along beaches looking for things that have been left behind by the receding tide. Once, as a teenager, I found a dead gannet on the strand line of the beach at Brora in north east of Scotland. I had never seen such a large bird up close. I picked it up, holding the wings out to the side. The smell was absolutely dreadful and, as I laid the bird down on its back, I could see that most of the underside had been eaten away by maggots. Some of them had fallen onto the pebbles and were wriggling out of sight. The head of the gannet looked intact and I wanted to keep the skull. I set about finding a tin can large enough to take the head and enough dry wood from the upper shore to make a small fire. Soon my gannet head was bobbing up and down in boiling sea water and later, I was able to take the cleaned skull home for further examination.

The inside surfaces of the upper and lower bill sheath were armed with hundreds of tiny, backward-pointing teeth. I ran my finger along the bill from front to back, and the surfaces seemed smooth, but doing it the other way was quite impossible. The sharp teeth tugged and pulled at my flesh. It was not difficult to understand how this would be a great boon for the bird. A fish seized by a diving gannet would slide easily between the bird's jaws but would not fall out before the gannet surfaced to swallow it. A gannet is well adapted for the job of catching fish. It is a superbly streamlined bird with no external nostrils, and air sacs on the face and chest to cushion it from the impact of diving vertically from a height of 30 metres at speeds of around 100 kilometres per hour.

But I had completely missed the bigger picture. No less well adapted were the hundreds of maggots that had been busily reanimating the dead gannet on the beach – transforming it in the space of a couple of weeks from one kind of flying creature into a different kind of flying creature altogether – many hundreds of them.

This chapter reflects that life on Earth depends on death and decomposition. Living things are made from the remains of dead things. There is a fixed amount of matter in the biosphere, so it follows that it must be used over and over again. Our bodies are made up largely of six elements, of which oxygen, carbon and hydrogen make up 92%. Every single atom that makes up you and me has been used before. The great British polymath Jacob Bronowski said, 'You will die but the carbon will not – its career does not end with you. It will return to the soil, and there a plant may take it up again in time, sending it once more on a cycle of plant and animal life.'

Imagine if I died on a solo wilderness adventure. It wouldn't be long before large animals came along to eat chunks of me to feed themselves and their offspring. A multitude of smaller creatures like ants, flies and beetles would also arrive to take their share of my remains. My bodily fluids would leak into the ground and be taken up by soil animals as well as plants. If we could actually follow all of the atoms that had made up me, we would discover that they would quickly become pretty widely spread around. At an atomic level we are immortal. It is a certainty that all the atoms in our bodies have been used millions of times before and will be used millions of times again before life on Earth is finally extinguished. This is not a nihilistic viewpoint – it merely describes how things are. The Earth is nothing more than a giant recycling plant, and a significant part of the process is carried out by insects.

The world is full of flies

You might think that blow flies are not that common, except perhaps in the warm summer months. Well, you're wrong. Variously known as carrion flies, bluebottles and green-bottles, these flies are pretty much everywhere most of the time – waiting to detect the faintest whiff of something rancid or rotten that will tell them that a good place to lay their eggs is not far away. The name 'blow fly' refers to meat being fly-blown – in other words, meat that has visible fly eggs on it. When filming high in the Himalayas, there were an unusually large number of bluebottles around our camp-site and, naturally, I wanted to know why. The flies showed me exactly where to look. I opened the flap of a small tent where the supplies of meat had been hung and saw that most of it already sported patches of fly eggs. It was well and truly fly-blown. Some of the eggs had already hatched and tiny maggots were visible. Eating a well-cooked curry made with this meat probably wouldn't have done me any harm at all, but that night, and for the rest of the time we were there, I went for the vegetarian option.

A really easy way to show just how widespread these flies are is to make a trap. Take a large, empty fizzy-drink bottle and cut off the base. Tape a piece of fine muslin over the hole. Now cut off the top off the bottle at the point where the curved section becomes parallel-sided. You will now have a funnel-shaped bit that you can turn round and stick back inside the bottle to form a kind of lobster pot. But, before you tape the edges to keep the funnel-shaped section in place, pop a fish head or a bit of meat inside the bottle. When it's all taped up, tie some string round the middle of your bottle trap and hang it up outside. I can almost guarantee that, especially in warm sunny weather, the flies will be arriving before you've even finished hanging it up. It might take a little longer in winter when it is cold and there are fewer flies about, but they'll find it soon enough. The only way the flies can get at the meat inside is through the screw-top hole and, as the wind blows through the bottle and out of the muslin-covered end, your trap will attract more and more flies.

When blow flies arrive, egg-laying will begin imme-diately, and you can watch the whole spectacle unfolding before your eyes. Flesh flies are rather different to blow flies in that they lay larvae, not eggs. By retaining their eggs and popping out newly hatched larvae, they are giving their young a head start in a highly competitive world. Several species of fly will arrive over the next few days and, eventually, you will also see small parasitic wasps arriving. These wasps might be attracted by the smell of decay, but they will also lay their own eggs inside the fly larvae, which are busily feeding on the decomposing meat inside. Over the course of a week or two, a whole microcosm of decomposition will unfold. When the fly larvae have eaten all they can and the remains become dry and shrivelled, several beetle species might come along and finish the job.

To really understand this process of decay, you've got to watch it happen. Years ago, in 2013, I presented a BBC programme called *After Life: The Strange Science of Decay*. It was a piece of televisual theatre that had never been attempted before and I'm very pleased to say that it won some awards.

The rot box

In a room in Edinburgh Zoo, a large sealed glass box was built. It contained a family kitchen with a small outdoor area. The kitchen was filled with all manner of fresh food. There was raw chicken and fresh fish left uncovered on the work surfaces, bowls of fruit and boxes of prepared sandwiches. Outside on the patio area was a suckling pig on a spit, a small plot with a kitchen garden and a compost heap. It really looked as if a family who had been preparing a big celebration party had been suddenly and mysteriously abducted. A half-empty wine glass on the worktop suggested that someone might have just gone to the toilet and would be back any minute. We heated the whole thing to about 25°C and left it all to rot. Fixed cameras in the ceiling recorded all the action continuously over eight weeks. Visitors to the zoo were able to walk round the outside of what became known as the 'rot box', and they could peer inside to see how things were getting along. The air coming into the box was sucked in from the outside and this gave plenty of opportunity for bacteria and fungal spores to enter – not that they weren't already present when the set was assembled.

But there was a problem with the flies. To solve this, I made up a couple of my bottle traps and collected about 100 or so flies from outside the building. I released the flies inside and watched them busily flying around their new home. Not surprisingly, the pig on the spit was one of the flies' first ports of call. In the course of a day or two, the flies laid many thousands of eggs on the slowly rotting fish, chicken and pig. To us, the smell of decaying meat is overpoweringly obnoxious and makes us want to retch, but after several hours of filming inside the box, I became less and less aware of it. If we had not evolved a sense of smell that reacted so strongly to decaying food, we might not have survived so well as a species. Even after showering and changing my clothes, people could still smell the whiff of decay as I walked past. This is because many of the molecules associated with decomposition – appropriately named organic compounds such as putrescine and cadaverine, resulting from the breakdown of amino acids – have a particular property: they are highly electrically charged.

It's a bit like when you rub a balloon on a woollen jumper and stick it to a wall. The molecules of the balloon pick up a charge that makes them stick to other materials they come into contact with. This is why the smell of decay can literally stick to our skin, hair and clothes. Leaving the rot box one night, I called a cab to go home for a hot shower. The driver wrinkled his nose and declared emphatically that I was 'fair minging' and that he didn't really want me stinking his cab out. I explained what I had been doing and the taxi driver gave me a withering look. He reached into the door pocket and fished out a can of deodorant – the one that shares a name with a genus of medium-sized wild cats (you know the one I mean). I gave myself a wee spray down, but I'm not sure it improved matters: I now smelled like a dead pig and an adolescent boy heading off on a hot date.

The time eventually came when I could cut open the pig carcass to see what had become of it. I took a sharp craft knife and cut through the dried skin. As I pulled it back, it was immediately clear that there was no meat left whatsoever – just bones and some fat remained. The pig had been entirely consumed by thousands and thousands of fly larvae. Many of them had already become adults and were now flying around the rot box looking in vain for something freshly dead to lay their own eggs on. In a piece of unscripted inspiration, I looked up at the new generation of flies filling the air around me and then turned to the camera. 'Who says pigs can't fly?'

Undertakers

In one corner of the rot box's outside area we had placed a small planter filled with soil. I was hoping that we might be able to film a pair of carrion beetles burying a dead rat. These beetles are also commonly known as sexton beetles, sextons being the church officials whose duties include looking after graveyards and digging graves. There are around 200 species of these industrious insect undertakers, and the reason they bury dead animals is so that they can lay eggs on them and raise their young. Many different insects and mammals are attracted to corpses, so if the beetles don't act fast, they will lose their prize. Equipped with exquisitely sensitive antennae, carrion beetles can detect a freshly dead animal, such as a bird or a mouse, from quite a distance – and they need to get there as quickly as possible, before flies lay their eggs or some other animal comes along and carries it off. Then there are other carrion beetles to worry about. At the height of the beetle breeding season, corpses are in high demand. A single male beetle finding a dead animal will try to attract a female by releasing a sexual pheromone. On smaller corpses, fights between pairs of beetles are common, but if the dead animal is large, several pairs of burying beetles will cooperate and use the resource communally. The beetles will even care for each other's young.

We had collected a couple of suitable beetles, a male and female from a local woodland. We popped them into the soil of the container and laid a dead rat on top. The miniature stage was set – low lights, camera, action. We left the bank of time-lapse cameras clicking away and came back the next morning. But the rat was still there, exactly where we had placed it. I racked my brains trying to think why our set-up had not worked. We reviewed the images and could see that the beetles hadn't even bothered to make an appearance. Something was clearly not to our beetle's taste and the only thing it could be was the rat. I couldn't really believe that our two beetles weren't able to bury a rat on their own, and I wondered if our rat wasn't as freshly dead as it might have been. The rats, which had been given to us by the zoo, were normally used as food for the pythons. We had to try again. This time, I requested a really fresh rat. At 7 p.m. the lights were dimmed as low as possible and the cameras were set.

The next morning, to our enormous joy, the rat had vanished. When we reviewed the footage, it was everything that we could possibly have wanted. The beetles began by digging a mine shaft at the rear end of the rat and gradually, the rat began to sink. Occasionally, the beetles would reappear and scurry around the corpse before returning to work. It looked as if the whole rat was being pulled at a slight slant underground by its tail. The last bits to disappear from view were its nose and whiskers. Once the rat was fully interred, our cameras were not able to see what was happening, but as we scrolled through the footage, some fur suddenly appeared above ground. The beetles had shaved the rat with their mandibles and carried most of the fur up to the surface. What happens next, although we could not film it, is that the beetles spread the skin of the shaved rat with secretions from their mouths and anal glands. These secretions have an antifungal and antibacterial action that keeps the meat from decaying too quickly. In the wild, the smell of decaying meat would soon attract a fox, and all their labour would be wasted.

The female beetle then lays her eggs in the soil around the rat, and the hatched larvae move inside the rat carcass, into what has become a comfortable nursery crypt, which the adults have lined with fur. The carrion beetle larvae can feed for themselves, but they also beg from their parents, who regurgitate some pre-digested food for them. This sort of parental care is quite unusual among insects, but I don't want you to get the wrong idea. The parent beetles may already have killed and eaten some of their own young to match the number of offspring to the size of the carcass they are about to consume. If there weren't enough resources to go around, the larvae would be smaller and this would mean smaller and less successful adult beetles.

A few weeks passed and it was time to exhume the remains of the rat. I was feeling nervous because, when you are filming stuff like this, no second takes are possible. The great thing about filming as things actually happen is that my excitement is absolutely genuine. If I had known in advance that it had all been a success, my reactions on camera would have been a play-act. I started with a small trowel and, about 25 centimetres down, I began to make out the spherical outline of the crypt. It was about the size of a large orange and inside I hoped I would find what was left of the rat. I used a small vacuum cleaner to carefully remove the soil around the fragile ball. As soon as it was free, I could pick it up and gently break it open on camera. It was spectacular. About 20 chunky beetle larvae could be seen and they had picked the rat's bones clean. Our two beetles had reared a healthy brood of young. The sequence of the actual burial process, which had been filmed over several hours and had taken many days to set up, would be speeded up to last only 30 magnificent seconds on screen.

A very nasty niff

The stench of decay, as I mentioned earlier, is utterly nauseating. There were times inside the 'rot box' when I retched and heaved. You cannot easily fake a reflex like this and make it look convincing on camera. Bizarrely, of all the food we had left to rot, one item seemed as fresh as the day we opened the packet. Eight weeks on and this particular foodstuff did not seem to have changed one little bit. No fungi had grown on it and no flies or beetles have gone anywhere near it. It was an iced birthday cake bought from a well-known supermarket chain. I cut a slice and sniffed it and could not even detect the faintest odour of bacterial decay. It does make me wonder what sort of preservatives are used and if we should really be eating them. The same could not be said of the whole raw chicken, now bloated and repellent, that we had left uncovered in the kitchen area. It was all I could do to be up close to it, but we wanted to use a portable analyser to sample the gases it was giving off.

I switched the machine on and thrust the sample tube into the rear end of the chicken. Unbelievably, the machine did not react at all. 'Is this thing working?' I asked the production team. I was assured that it had been working.

Now, by way of explanation of what I am about to tell you, one of the great advantages of working inside a stinking sealed box is that flatulence is no longer a problem, and personal bowel gas can be released at pretty much any time with impunity. On the spur of the moment, it occurred to me that a good way of testing the gas analyser was – yes, you've guessed it. I held the machine behind me and let out the teensiest fart I could manage. Immediately, the lights on the machine started flashing madly and the air was filled with the noise of a high-pitched siren. Now I had proof that the gas analyser was working perfectly, but I also realised that my wind contained considerably more hydrogen sulphide and ammonia than a six-week-old chicken carcass.

Team work

Next to the chicken were two packs of hamburgers. One pack had been torn open to let flies get to the contents, while the other pack remained sealed. Of course, no fly larvae were seen in the sealed burger pack, but it had bulged alarmingly nevertheless, as the few bacteria that were present when the pack was sealed had now multiplied. But our time-lapse cameras revealed some interesting behaviour in the opened pack. Inside, two identical hamburgers lay side by side – but all the maggots were feasting on one of them and were ignoring the other burger. Speeded up, they scurried about, reducing the burger to a sloppy mess. Then, all of a sudden, they migrated en masse to the other burger. It looked remark-ably like the maggots were hunting in a pack.

In fact, there is good reason for this communal feeding behaviour. A single maggot on its own would not do nearly as well as it would being part of a larger group, in which all of their digestive enzymes can be pooled. Our thermal cameras also revealed that the temperature had increased significantly inside the feeding ball of maggots. Indeed, it can get so hot inside a maggot ball that larvae on the inside must occasionally move to the outside of the ball to cool off. As a result of the increased temperature, the flies develop at a much faster rate. We also had at our disposal an incredible camera that could film live subjects at tremendously high magnification. This allowed me to take a closer look at the anatomy of a maggot and show how perfectly adapted they are for turning dead flesh into flies.

Once the first-stage fly larva has hatched out of its egg, it will feed on fluids and burrow its way inside the corpse. In a day or two it will moult to become a larger, second-stage larva. When it reaches the third and last larval stage, it will be feeding with many others in a mass. The larvae look similar in all three stages, except that they have increased in size from about 2 millimetres to nearly 20 millimetres. The head end is slightly narrower than the rear end and they are elongate, cream-coloured and soft-bodied. At the head end there is a pair of toughened structures called mouth hooks that rip and tear into the food. The body contains a pair of large salivary glands that pump out enzymes, reducing the food to a soup-like mush. The segmented body is muscular, with bands of raised welts that act like the spiked shoes of athletes and help the maggot wriggle through the sloppy food. At the broader rear end are two noticeable dark structures. These are the openings of the tracheal system and, being located at the back end, allow the maggots to feed almost continuously.

Watching them as they fed, I couldn't help but be impressed by the sheer efficiency of the whole process. When the larvae are fully grown, they wander off to find a suitable place to pupate, usually in the soil close by. The pupa is formed inside the skin of the last larval stage and, over the course of around 10 days, the remarkable transformation to adult fly takes place. If we had provided food at regular intervals, it wouldn't have been too long before the entire rot box would have been knee-deep in dead flies. As it was, the place began to gradually dry out, and it was later that I discovered several thousand flies crammed inside a half-empty bottle of red wine. I hope they died happy.

Medicine and murder

One of the most well-known fly species is the Common Greenbottle. Found pretty much all over the world, this rather beautiful blow fly has a characteristic metallic green appearance and is about 10 to 15 millimetres long. You will see the adults basking in the sun and feeding on nectar at flat-topped flowers such as cow parsley. But you will also find them feeding on wet faeces and on dead animals, where the females lay their eggs. We humans like to categorise things. We need to put things in pigeonholes. And for many of us, insects are either nice or 'useful', such as butterflies and bees, or nasty and disgusting, such as hornets and blow flies. Of course, this view is simplistic and consequently rather unhelpful.

The Common Greenbottle is a wonderful example of an insect whose life cycle and habits may at first seem revolting, but which is useful to us nonetheless.

Blow flies are the primary agents in the decomposition of animal remains, but sometimes they do not confine themselves to dead tissue. Greenbottles cause fly strike in sheep – a debilitating and potentially fatal condition where maggots start eating the sheep alive. Greenbottles are attracted to fresh carrion and have evolved to be the first at the scene, where they can complete their life cycle – or at least get it well under way before anything else comes along.

In the case of fly strike, the greenbottle females are attracted to the faeces- and urine-soaked fleece at the rear end of a sheep. Here, they lay their eggs, and the hatched larvae burrow down towards the skin. Living skin would normally be a reasonable barrier, but the mass action of lots of little maggots scraping away will eventually break the skin's surface. The young maggots then feed on fluids coming from the weeping wound and will eventually start to eat live tissue. In many breeds with heavy fleeces, it is common for the sheep to have their rear ends shorn to minimise the build-up of soiled wool, which would attract blow flies. By breeding sheep with unnaturally luxuriant wool for fashion industry purposes, we have provided a great opportunity for the ever-adaptable blow fly. And now urban areas have welcomed them by giving them yet another abundant resource to colonise – wheelie bins and food-waste bins.

Blow flies like the Common Greenbottle, *Lucilia sericata*, have also become the stars of murder fiction and television crime dramas. As a fresh corpse decays, it will attract waves of various species of insects. Because the development time of different species at different temperatures is quite predictable, flies especially have become a hugely important part of forensic science. We are all familiar with the scene in which the police enter a building where they will discover a corpse. The door to the room where the body lies is closed, but there is a terrible smell and the sound of buzzing flies coming from inside. Blow flies will find a dead body quickly and, in warm conditions, a new generation of flies can be produced in just under three weeks. But how many flies might you expect to see in the room if they were unable to escape? I once reared 1,700 flies from a blackbird killed by my cat. A rat might provide enough sustenance to produce 4,000 blow flies. It has been estimated that fly larvae can consume 60% of a human corpse in a week, so we must be talking about a substantially large number of flies indeed.

Flies are particularly useful in estimating the time of death in murder cases. But to get to the point where this information can be used reliably in a court of law, you need to study and become intimately familiar with the behaviour and physiology of the insects involved. The best way to do that is to watch what happens when a dead body decays. Body farms where this kind of work can be done exist in the United States, Canada and Australia; at these farms, human corpses are left to decay in diverse ways, to show how decomposition takes place under varying conditions.

In the UK, forensic scientists have to content themselves with domestic pigs instead.

I visited a 12-acre site in the northwest of England, where, for more than a decade, this essential research has been carried out. Surrounded by stone walls and other farms, an old horse pasture has been used for the experimental study of how bodies decompose. Nothing prepared me for the sight of a gibbet with a row of pigs strung up by the neck, simulating hanged humans. Not just hanged pigs, but buried pigs, drowned pigs, pigs in boxes and pigs in blankets (quite literally) – standing in for victims murdered in every conceivable manner. Pigs are used because they are in many ways anatomically and physiologically similar to humans. Their heart is about the same size as ours and their cardiovascular system suffers from the same complaints, such as hardening of the arteries and heart attacks. Their skin is also a good match for ours, so, although not quite perfect, dead pigs are great substitutes for human corpses.

The moment the heart stops beating and blood stops flowing in an animal, the process of cell death begins. Able to detect the presence of a dead animal from more than a mile away, blow flies arrive quickly and lay their eggs in easily accessible wet areas, such as open wounds and around the eyes, nose and mouth. It was once believed that blow flies could detect death just before it happened. But don't be too alarmed if you see a few greenbottles following you – it's more likely that you've just trodden in dog shit.

As cell decay progresses, each cell membrane splits and the enzymes inside are released and will begin to break down other cells. Bacteria from the gut then start to feed on the protein-rich fluids, releasing the gases that cause the animal to swell up or bloat. The body is now well and truly breaking down, and splits in the skin may appear, allowing even more flies to lay their eggs. This stage, known as active decay, is when most of the mass of the corpse is lost due to the feeding of maggots and the leakage of bodily fluids. Blow flies can consume the body of a pig (or a human) in just a few weeks – larger animals might take a little longer. And speaking of large animals, what about the largest animals that ever lived – the dinosaurs? Would their corpses have been bloated and writhing with maggots after they died? It would have been a spectacular sight – and smell – for sure. But it never happened. The group of flies that includes blow flies and flesh flies are relatively modern, and only appeared after the mass extinction event of around 60 million years ago that brought about the demise of the giant reptiles. So what was removing the staggering amount of dead flesh before flies came on the scene? It seems that there would have been so many scavenging vertebrates around that nothing much would have been wasted. What small bits and pieces were left lying around would have been nibbled and gnawed by beetles of various kinds.

The scientists working on this macabre pig farm during my visit seemed inured to the daily assault on their sense of smell as they went about the serious business of taking samples and recording data. This data is important because decomposition is affected by so many factors. A body exposed to the elements attracts many more insects and decays much more quickly than if it was buried underground. How much clothing is present makes a difference too, but the most important factor of all is temperature, and to a smaller degree, humidity. The hotter it is, the faster the development of flies. Research like this helps to improve the accuracy when forensic scientists are called on to establish a minimum post-mortem time in a murder case. It's more or less a matter of looking at the state of development of the flies present on a corpse and counting the days backwards. The research can also shed light on whether or not corpses have been moved after death and whether there is any indication that death resulted, not from natural causes or an accident, but at the hands of someone else. Flies don't lie – oh, that's rather a good book title, but I imagine some crime writer may have already used it. As I left for home, and a long soak in a hot bath, I kept wondering how many murderers might still be walking free if it were not for flies telling tales.

But the maggots that feast on the corpses of murder victims and burrow into the backsides of sheep are the exact same ones that have been saving human lives for centuries. The use of blow fly maggots to clean up infected wounds is an ancient technique. It is known that the ancient Mayan civilisation of Mesoamerica used maggot dressings to treat superficial wounds, and there are historical records of many other places around the world where humans have used maggots in this way. From the sixteenth century onwards, several military surgeons observed that soldiers, some with gravely serious wounds, recovered in cases where maggots were present. They noticed that maggots improved their chances of survival and observed that the wounds were cleaner and healed faster than would have been expected. Despite it being rather counterintuitive, battlefield medics would sometimes allow wounds to become maggot-infested – and would even introduce maggots on purpose. In many cases the results were impressive, and contemporary accounts from the American Civil War leave no doubt that the treatment could be effective. One medical officer wrote: 'Maggots … in a single day would clean a wound much better than any agents we had at our command … I am sure I saved many lives by their use.' The practice of maggot therapy, or 'biosurgery', as it is also known, became a more recognised treatment for wounds after the First World War, when experience proved once again that lives could be saved.

By the 1930s and 1940s, maggot therapy was offered in many hospitals, and there was a reliable method for rearing the maggots in sterile conditions to avoid the risk of accidental infection. The larvae of the same species that cause fly strike in sheep have become the medical maggots of choice – and the advantages of using maggots to clean up wounds are numerous. In a typical case, a small number of maggots are placed in a wound and left in place for a few days, after which time they are removed. The maggots are able to remove dead material and debris far more carefully and precisely than even the most skilled surgeon using the finest scalpel. As they feed, the maggots produce tissue-dissolving enzymes and other secretions that have an antibacterial action, so that the number of potentially dangerous bacteria present in the wound is greatly reduced. The advent of antibiotics meant that maggot therapy became much less popular than it once was, but now that many bacteria have become resistant to multiple antibiotics, maggot therapy is having something of a renaissance. For example, it has been shown that maggot secretions are active against certain dangerous bacteria that have become a serious problem and that cause infections that are challenging to treat.

DR ERICA MCALISTER

The flies have it!

Professional entomologists are rarely generalists. Most focus their attentions on one insect group to the virtual exclusion of all others. True flies belong to the order Diptera, and they have a veritable champion in Dr Erica McAlister, a senior curator at the Natural History Museum in London. In social media circles Dr McAlister is affectionately known as 'the fly girl'.

I met up with Erica at the museum where she was photographing a fly specimen that was to be sent out to a researcher as a digital loan. I asked why she was sending a photograph rather than the actual fly, as used to be common practice. She explained:

'Our specimens are the name-bearing specimens for every species. The entomology collections at the Natural History Museum represent about half to three quarters of the described species of life. These are the specimens that were collected when it was discovered they were a new species. They were sent to a museum to be kept in perpetuity.

'People ask, "Why are you keeping these species and why do you still also need to kill specimens?" None of us do this lightly, but we need to be able to look at all of the possible different structures. We then make a detailed description and diagram for future generations to refer to. In the eighteenth century, zoologists Carl Linnaeus and Johann Christian Fabricius produced no images, and their species descriptions were shocking. The house fly was described as 'black fly with bristles', narrowing it down to about 7,000 species. As time goes on, we've gotten better at describing, and looking at their DNA as well as their morphology.'

When I was active in museums, it was terrifying when you had to send material. I used to have to send type specimens round the world, and you had no idea if the person would look after them. But technology has moved with the times.

'Our automated stacking system takes hundreds of images. They get compressed together to make a three-dimensional or in-depth image of the fly.

'We have a responsibility to maintain and enhance this collection and ensure all the data is now available for everyone to access it. The loan I am currently working on will not only go to the person who requested it, but it also gets put online, available to anyone who wants to use it. This frees up research budgets massively because depart-ments will no longer have to pay so much money to visit specific collections as they will have such high-quality images online.

'It's crucial that we retain all holotypes [a single physical example of a species], and this is especially important when you start understanding medical species. With mosquitoes, the most important to research were the complex species of *Anopheles gambiae* and *Anopheles funestus*. Using the holo-types, we can look back and analyse these historic specimens for their DNA. We have the technology to confirm which species are which. If we didn't have the original specimen, that wouldn't be possible, so it's imperative we keep these specimens in the best way.'

Erica showed me some specimens preserved in a jar of ethanol. The label said 'Maggot from the bodies of Ruxton's victims 1935. Moffat Murders.' I know the Scottish border town of Moffat really well, but I hadn't heard of this.

'The famous Ruxton maggots were used in the first forensic entomology case in the UK. A doctor killed his wife and housemaid and the maggots were part of the case against him, as they provided a timeframe. By knowing how old the maggots were, they could ascertain the *minimum post-mortem interval* – the earliest the bodies could have been deposited there. This evidence was one of the many reasons the man was hanged for his deadly deeds.'

Erica also showed me all manner of other weird and wonderful creatures preserved in glass jars for study: camel spiders, giant centipedes, whip spiders.

'When you look at the structure in arthropods, the variety is amazing – it's all so different, and I love that. Once you've seen one primate, you've seen them all.'

We move to another part of the museum where there were serried ranks of steel cabinets on moveable racking. We open a cabinet containing asilids, or robber flies. And look at drawer after drawer of neatly pinned flies.

'We have five floors for entomology and two floors for botany. Somebody has already gone through and put specimen numbers on them all – we have a spreadsheet of all this. What I'm trying to do is go through all the British collections and barcode them. When the spreadsheet goes online, we'll suddenly have access to all this information about when these specimens were collected.'

There are only four staff in the fly section. I asked Erica if that was enough.

'It's not, considering how many species there are. There are 165,000 described species so far. In the UK, there's over 7,000 – but we are luckier than some. It's a really important collection and one that we add to all the time.'

I wondered when she first became interested in insects, and whether it was always flies that captured her attention.

'I try to find an exact moment where my interest started. I wasn't like Darwin, stuffing them in my pockets as a small child, but I was a very accident-prone kid, constantly falling over and spending a lot of time on the ground. My dad gave me a microscope very early on. I'd collect fleas off cats and things like that. I did keep some dead mammals under my bed for a while, as you do, but it was in the name of science. I've always been fascinated by the ecology and love knowing what's going on. There's something beautiful in asking, "What are you?"'

She recalled watching a fly emerging from its pupal case in her garden.

'The fly basically reorganised most of its body tissue from a maggot into this amazing metallic beast. Why would I study anything else? Flies get everywhere and do everything. They're weird, crazy and fun. There are marine flies that are the best explorers, medics and dustmen. They're everything and I love them.'

Flies may well be the unsung heroes of the insect world, and if anyone can act as their cheerleader, it's Erica McAlister. Her enthusiasm is totally infectious and, as I leave the museum, I resolve to look at flies much more in the future.

Nothing is safe

When I started work in the Oxford University Museum of Natural History, one of my major tasks was to look after the extensive collections of insects. I had seen the extent of them during a tour of the department just before my initial job interview. It was daunting, to say the least. Some of the collection was housed in purpose-built entomological cabinets, but there was a substantial amount of material pinned into double-sided storage boxes of various ages and conditions. Some of the boxes were definitely antiques and a few even looked as if they predated the opening of the museum building in 1859. Several rooms were filled with piles of these boxes arranged in precariously teetering stacks. I picked up one from an unusually low pile and opened it. A flurry of moth wing scales billowed up into the air. Many of the pins that once held dried insects were bare. The bottom of the box was covered with a fine dark grey dust and the broken remains of many specimens that had been collected and carefully pinned for posterity more than a hundred years before. I knew I had a battle on my hands. My enemy was a legion of small beetle larvae that had a voracious appetite for all sorts of dried organic material and were more than capable of destroying an entire museum insect collection.

Variously known as skin, carpet or museum beetles, their very presence threatened to turn Oxford's stockpile of scientific specimens to dust. As adults, museum beetles are only a few millimetres long and have a characteristic rounded shape. Up close, they are rather attractive, with patterns of coloured scales. The adults feed on pollen, but the larvae – which are thickly covered in long hairs, giving them the common name woolly bears – are scavengers. As such, they are an important part of the end stages of decomposition, but you really don't want them anywhere near your woollens, hides, furs, feathers or insect collections. I have visited many houses where the horrified owners have moved an item of furniture only to find that the carpet beneath has been completely destroyed. Of course, they only eat natural fibres, so if your carpet is made from acrylic, polyester or some other synthetic fibre, you don't have to worry. If you see these beetles on the inside of your windows, it means they have bred somewhere in your house and are now trying to get out. You might overlook one or two, but if you see a dozen or more in the space of a few days, you may have a problem. My maternal grandmother, on learning that I was moving south to continue my education, was quick to impress on me that my kilt, now rarely worn, would soon fall prey to moths and other wee beasties.

Mothballed

In days gone by, all manner of noxious chemicals were used to control skin beetles. Insect collections used to absolutely reek of naphthalene, an organic compound made from the distillation of coal tar. It is a white crystalline solid with a characteristically pungent odour – think of old-fashioned mothballs. Now considered to be carcinogenic, naphthalene has not been used in mothballs or in museums for more than 40 years, and no one today has the unpleasant job of topping up the specially made cavities in insect drawers, but the residual smell still lingers in the woodwork. Like decay, the smell of naphthalene clings to your clothes and your hair. When I worked on my PhD at the Natural History Museum in London in the early 1970s, I remember taking the rush-hour Underground from South Kensington. I could tell immediately if a fellow entomologist was in the same carriage as me.

Oddly enough, some museum departments rather like skin beetles, as they are used to clean vertebrate skeletons. With large animals you can remove most of the flesh and guts beforehand, but with small creatures such as birds, bats or fish, wielding a scalpel might damage some of the fine bones – this is a job best left to the experts. When the specimen has been dried out, the beetles and their larvae are then sprinkled on it and left in a sealed container. The beetle larvae will eat all the dried flesh, sinew and gristle, and are able to get inside the skull and the smallest of spaces to eat whatever is available. Sometimes tiny bones get lost; to avoid this it is best to put smaller creatures inside a mesh box that allows the larvae in but nothing else out. Once the beetle larvae have done their work, you simply have to retrieve your specimen, degrease it, and perhaps bleach the bones to make them nice and white and voila! You've got the most perfect bone preparation you could ever wish for. I wish I'd known about this when I was a teenager, as my small collection of animal skulls was prepared by boiling dead animals in a pot. The smell was always pretty grim.

I once visited a museum (I won't say where, to save the curators' blushes) in which the entomology department was plagued with skin beetles. They were everywhere. It didn't take long to find out why. On the flat roof of the museum, the zoology curators would routinely lay out whole animal carcasses in the hot sun. The flies and their armies of maggots took the biggest share. Weeks later, the remains were swarming with skin beetle larvae busily cleaning up the bones and raining down through cracks in the building to the entomological cabinets two floors below, where they would find more to eat. Cartilage or cuticle – it makes no difference to skin beetles. Of course, bacteria and fungi are important agents of decay, but when it comes to bulk removal of material, insects are in a league of their own.

So far I have been talking about the decomposition and recycling of dead animals, but don't forget that plants make up more than 80% of the total global biomass. Plants store carbon and produce oxygen, both of which are vital to life on Earth, but they also lock up a huge amount of energy. A lot of that energy is in the form of an organic polymer called cellulose, the material that makes the structure of plant cell walls. There is a vast amount of cellulose available locked up in living and dead vegetation, and one group of insects has evolved to take full advantage of it. Termites are descended from cockroaches and are generally confined to warmer parts of the world where they decompose plants and dead wood (including paper of all kinds), as well as the organic matter in soil and animal faeces. Cellulose is normally hard to digest, but termites can do it because of symbiotic bacteria and single-called organisms they have in their guts. One of the unfortunate side effects of all the work that termites do in recycling dead wood is the production of methane, which contributes to global warming.

But before we start blaming farting termites for global warming, I reckon we are responsible for the production of a lot more greenhouse gases than they are. In any case, termites are, without doubt, the major ecosystem engineers in tropical regions. They are responsible for the recycling of huge amounts of nutrients and affect the formation and structure of soils over impressively large areas. More than this, having converted indigestible material into juicy termite flesh, and despite their best efforts to avoid being eaten, they support a rich community of animals that feast on them. Many arthropods, such as ants, spiders and centipedes, and vertebrates like reptiles, amphibians, birds and mammals feed on them almost exclusively.

Remember the aardwolf I mentioned at the beginning of this book, the one that disturbed my savannah stargazing? It has a long sticky tongue to gather termites, and primates such as chimpanzees use sticks as tools to fish termites out of their nests. But there's safety in numbers – there are, after all, an awful lot of termites. One highly specialist termite-eater is a species of lacewing whose larvae wave the rear end of their abdomens close to the heads of their termite prey and emit a volatile substance that paralyses them. One whiff from the lacewing larva's anus can even paralyse several termites at once. Once its meal is safely gassed, the lacewing larva eats the termites in peace and quiet. Of course, it was not long before the popular press got hold of the story: 'silent but deadly' and 'fatal farts' featured prominently in the reporting.

Being now much closer to the end of my life than the beginning, I do think about what I'd like to happen to my remains. I am rather horrified at the huge amount of energy it takes to cremate a human, so that's out. I'd really rather like a Tibetan sky burial, where my body would be left on a mountain for scavengers, insects and the elements to deal with, but this would be logistically problematic and expensive. I could give my body to a medical school so that students could hone their dissection skills on me or to a body farm, if there were any in the UK. Burial seems most likely – I do like a nice graveyard as long as it's a bit overgrown and there are lots of insects around. I will not be reincarnated as such, but the atoms I have been using during my lifetime will end up in many plants and animals.

By now you will hopefully realise that insects are pretty important in the great scheme of things. They are the lifeblood of global ecosystems. If you still think we could get along without them, the next chapter might change your mind.

Chapter 7

WHAT HAVE INSECTS EVER DONE FOR US?

One day, when I was working in Tanzania, my eye was drawn to a large and colourful shield bug resting on the top of a flower. I ran over to look at it, but my African colleagues stayed where they were and began talking quietly among themselves. I had stood at the flower for less than a minute when I became aware of things moving about under my trousers. Seconds later, sharp stabbing pains in my nether regions caused me to beat a hasty retreat. It was only when I took my clothes off that I realised I had been standing slap bang in the middle of a column of marching driver ants. 'Why didn't you warn me?' I asked while picking dozens of ferocious soldier ants from my skin and underpants. I didn't get a reply, as they were already helpless with laughter. This experience would have put many people off ants for life. But not me. It only showed what incredible creatures they are – just not when they're in your pants.

I have always felt the need to tell people about how wonderful insects are, and for the first part of my career, it was pretty straightforward. Undergraduates are, by and large, a receptive and captive audience. Today, more than half of the world's human population live in urban areas where the natural world is, to a large degree, out of sight – but it must never be out of mind. More than ever, we need to know about the creatures that make the world, our world, work.

You know that bit in the film *Life of Brian* (1979) where Reg, the leader of the People's Front of Judea, played by John Cleese, asks, 'What have the Romans ever done for us?' Gradually, his followers mention a number of hugely beneficial things, from roads, aqueducts and education to sanitation, public order, irrigation and wine. Reg accepts rather testily that, yes these have been generally been good things, but he keeps asking his audience, 'Apart from roads, aqueducts, education, sanitation, public order, irrigation and wine – what have the Romans ever done for us?'

This is exactly the purpose of this chapter – just what have insects ever done for us?

If I asked an audience, I'm pretty sure I would get a range of answers. I hope some would point out their crucial role as pollinators or recyclers or as the food for most species on Earth. But it wouldn't be much of a surprise if some people dwelled more on the dark side of insects. Insects are, after all, the vectors for a large number of disease-causing organisms that affect plants and animals and that have killed millions upon millions of humans.

Plague

Some insects have even changed the course of human history. Between 1346 and 1353 a great plague swept over Eurasia and North Africa, bringing hysteria and death. At the time, some people believed it was a pestilence sent by God as a punishment for all our wrongdoings. Some astronomers had also seen a strange conjunction of three planets in the heavens and thought that this event had somehow brought about a malignant miasma which was carried in the air. The plague, or the Black Death, as it became known, was the worst pandemic in our recorded history and as many as 200 million people may have died from it worldwide. In most towns and cities, at least half of the population succumbed fairly quickly. The rapid spread of the disease, accelerated by active trade routes, must have been absolutely terrifying. Charlatans and quacks were quick to offer cures, but they didn't last long enough to enjoy their ill-gotten gains – and with few people to bury the dead, corpses piled up in ditches and in hastily dug pits. Dogs dragged the remains from shallow graves and ate them. People turned in desperation to doctors and priests in the hope of some protection, even salvation. Poorer people that lived crowded together in urban areas fared worse, but even so, the plague was killing everyone, no matter their social standing. Neither prayers and potions nor your position in society made any difference at all. The situation was so terrible that people must have thought, quite reasonably, that the end of the world had come.

Now we know that the plague is caused by a bacterium that is carried in the gut of a flea after it has fed on an infected rodent such as a rat. Fleas are recognisable to us; these small, wingless insects are no more than 3 millimetres long and are distinctively flattened from side to side with a characteristic ability to jump. The vast majority of flea species are external parasites feeding on the blood of land mammals, with the remainder feeding on bird species. Fleas are found wherever there are suitable hosts, even in Arctic and Antarctic regions. The only mammals that escape their attentions are aquatic or have incredibly thick skins. I'd hate to meet a flea that could puncture the hide of an elephant or a rhino.

I once shared a house with a biologist who had two pet cats. The house was pretty untidy and grubby, and it was clear that the cats were infested with fleas. Fleas drop their eggs into the bedding or nest material of their host animal and here, the tiny, worm-like larvae feed on particles of organic matter, including dried blood and the excrement of the adult fleas. When fully grown, the larvae pupate inside a silken cocoon camouflaged with minute bits of debris. If no suitable host animals are around, the adult fleas can stay dormant – people moving into a house where pets have lived previously may be attacked by thousands of starving fleas that have matured in the absence of anything suitable to bite. I tried hard to keep my bedroom flea-free and, on returning from the bathroom after a shower, had to inspect my legs carefully. Cat fleas can jump to a height of 30 centimetres hundreds of times an hour for several days to find their next victim – and I had to find a new place to stay.

The plague bacterium primarily infects rodents but can be passed to humans when the fleas' hosts die and the fleas are forced to seek blood elsewhere. A flea feeding on an infected brown or black rat might pick up the bacteria in a blood meal. The bacteria multiply in the midgut of the flea, and in some cases may completely block the flea's digestive tract. In this case, when the flea attempts to feed on the next host, it cannot pump up the blood into its blocked gut. When it stops trying to feed, pressure and elastic forces cause the blood in the oesophagus, which is now mixed with some of the plague bacteria, to be ejected back into the host. In this way, a flea with a blocked gut can infect many hosts before it eventually dies of starvation.

There are three distinct forms of plague: pneumonic plague, septicaemic plague and bubonic plague, distinguished by the route by which the bacterium infects the victim. Pneumonic plague affects the lungs and results from the inhalation of infected airborne droplets, while septicaemic plague infects the bloodstream, and bubonic plague infects the lymphatic system, giving rise to greatly swollen lymph nodes in the groin, neck and armpits. These swellings, called buboes, can rupture, releasing blood, pus and bacteria in large amounts. Plague victims would suffer fevers, nausea, vomiting, agonising cramps and seizures, delirium, coma and eventually organ failure and death. The extremities of the body appeared black as the skin tissue died. There is no doubt about it – plague is a truly ghastly disease and an incredibly painful and unpleasant way to die.

The Black Death brought about a series of religious, social and economic upheavals far greater than those resulting from war, and these had far-reaching effects on the course of European history. People's faith in the church had been seriously shaken, and their confidence that those in authority actually knew what they were doing was at an all-time low. Survivors who could still work on the land were able to choose where they worked and for whom and, as a result, wages and standards of living slowly improved. It was nearly two centuries before the population of Europe recovered, but things were never going to be the same again. Plague recurred occasionally in Europe until the nineteenth century, and small outbreaks in parts of India, Asia, Africa and South America took place in the twentieth century – and can still occur today under certain conditions. But before we demonise the flea, remember that they were unwitting vectors, and as much a victim of the plague bacterium as rats and humans.

Blood-suckers

You cannot ignore the telltale whine of a mosquito. It is an unmistakable and unsettling high-pitched buzz that means trouble. I have lost count of the number of times I have lain under a bed net in various tropical countries listening to that disquieting sound. Jungle birds, frogs and monkeys can make as much noise as they like, and it will not keep me awake – but one single mosquito flying in the shadows is another thing altogether. I know that until I have inspected every single square centimetre of my mosquito net for holes and tears, no matter how small, and made sure the net is properly tucked in, I will not be able to drift off to sleep. The mouthparts of a mosquito are made up of several slender, elongate stylets which are gently sawed and pushed through skin and flesh towards a blood vessel. Saliva, that temporarily numbs the pain and stops the blood from coagulating, is pumped through the needle-like structure and the blood meal is sucked back up.

There is no doubt that the most dangerous insect on Earth is the mosquito. There are several thousand species of these delicate, slender-legged flies worldwide, and they are well-known blood-feeders, attacking mammals, birds, reptiles and amphibians, as well as a few invertebrates for good measure. Although both sexes feed on nectar as an energy source, it is only the female mosquitoes, who need to grow and mature hundreds of eggs, that need blood meals. Mosquito larvae are aquatic and can develop in any still or slow-moving water, including marshes, mangroves, swamps and even stagnant or brackish pools. Almost anywhere rainwater collects – be it a natural tree hollow, a tin can or a discarded tyre – can act as a breeding ground.

Mosquito bites can be tremendously itchy, but much worse than that, your blood may become infected with any one of a number of disease-causing organisms, such as yellow fever virus, West Nile virus, dengue virus and Zika virus. The worst mosquitoes of all are some species in the genus *Anopheles* that act as vectors for the organisms that cause malaria. In the late 1800s, the discovery of the malarial parasite, a single-celled organism called *Plasmodium*, and the role played by the mosquito in the life cycle of the parasite, made prevention a possibility. The disease can be treated with an extract from the bark of the cinchona tree, from which the powerful plant compound quinine was eventually extracted.

Quinine interferes with the ability of the *Plasmodium* parasite to metabolise haemoglobin, and today there are several more effective anti-malarial drugs. In the fight against malaria, the breeding grounds of mosquitoes were drained or covered with oil to suffocate the larvae, and huge amounts of insecticides were used to kill the adults. The development of DDT and its extensive use after the Second World War helped rid many countries of malaria, but it was not long before the mosquitoes developed a degree of resistance, and the use of some insecticides was banned because of their persistence in the environment and the harmful effects they had on other organisms. Mosquito control programmes and the use of bed nets have been successful in reducing the number of malaria cases, but the disease will never be properly dealt with until effective vaccines become widely available.

Malaria has probably killed around 5% of all humans that have ever lived, and the impact of malaria continues to be huge. It is estimated that nearly half of the human population is currently at risk, with the vast majority of cases occurring in sub-Saharan Africa. Globally, there are more than 400,000 deaths per year, and children under the age of 5 make up well over half the total. The amount of money spent on malaria research is far smaller than is spent on many non-fatal illnesses, and I have to wonder whether malaria might have been eradicated by now had it been a disease of the northern hemisphere. However, even as I write these words, a ground-breaking malaria vaccine is showing great promise and may save tens of thousand of young lives in the coming years.

Bugs on the battlefield

More combatants in warfare have died of insect-borne diseases than were ever killed on the field of battle. Diseases such as epidemic (or louse-borne) typhus, associated with insanitary and over-crowded conditions, has contributed significantly to the outcome of many wars. The causative organism is a bacterium that is passed from person to person by infected lice. The bacteria grow inside the gut of the lice and are passed out in their faeces. The bites of these small, wingless insects itch, and scratching them only serves to rub the typhus bacteria into the fresh puncture wounds. The mortality rate is around 40%, and millions of humans died before antibiotics became available. Ironically, the advent of insect control through pesticides along with more powerful, remotely operated weapons has enabled us to kill ourselves with far more efficiency today than insects ever did.

Insects don't need to transmit diseases to be a major pain in the neck – or wherever else they bite you. Blood-feeding flies are almost certainly a major reason why some parts of the world remain thinly populated. I would like to take this opportunity to record a debt of gratitude to the Highland midge for keeping large parts of Scotland relatively pristine. Without the ferocious habits of these tiny flies, it is more than likely that large parts of my native country would already have been given over to holiday homes and golf courses with the loss of much natural habitat. The reason midges are so hard to get rid of is that the larvae develop in boggy areas, and although adult midges do not fly far from where they bred, enormous parts of Scotland are very wet indeed. Many years ago I travelled by train on the West Highland line to Mallaig and got into a conversation about midges with an elderly passenger. 'Aye, they're bad,' he assured me – and then told me a story about a man from his village who had been making a nuisance of himself with the local lassies. They dragged him from the pub one night, took him to a midge-infested location, removed his shirt and tied him to tree. 'What happened to him?' I asked. The old man rolled his eyes and, with great relish, said, 'The next morning when we went to untie him – he was nearly insane.'

Good grub

But enough of the dark side – it's time I turned to some of the many positive aspects of insects. I have already mentioned that insects are the food of the world. Many vertebrates depend on them exclusively and because of this, some animals can be tricked into trying to eat something that is not quite as it seems. The use of artificial insects fashioned from feather, hair, thread and synthetic materials to imitate real insects on or below the water surface is the basis of fly fishing. While the art of tying a convincing artificial fly is important, the key to successful fly fishing is a good knowledge of the behaviour of the fish and the prey insects on which they might be feeding. As a result of an urgent request from a publisher who had been suddenly let down by a manuscript, Steve Simpson, my colleague at Oxford, and I wrote a slim volume called *The Right Fly* (2002). It was a tight deadline, but during the Christmas holidays we carved out some time, and a few weeks later, our text was completed. Steve, a keen fly fisherman, wrote sections on fishing and fish behaviour, and I wrote up a short compendium about the biology of the sorts of insects trout liked to eat. It was great fun. When it was published one of the reviews stated that it had been written by two 'Oxford biologists with a passion for fly fishing'. I had never held a fly fishing rod in my life, let alone used one, and Steve decided that this should be put right without delay.

We set off to a private lake in Oxfordshire, where Steve gave me a quick tutorial and a demonstration of the basics. It all seemed pretty straightforward – an expert always makes it look easy – and Steve left me to have a practice on my own while he, sensibly, as it turned out, moved around the margin of the lake to a spot where he thought he had seen some trout rising. I started to cast and it was not long before I'd caught something quite large – my leg. The hook had buried itself through my trousers, high into the inside of my right thigh. 'Steve,' I called out, 'I think I've caught something.' Steve came over and knelt down in front of me to remove the hook by pushing it all the way through my flesh before cutting the barb off. He told me later that the local gamekeeper had been watching and muttered, 'You Oxford chaps,' as he passed by. I've never gone fishing again.

Insect-eating is an ancient habit. Reliable, abundant and easy to gather, insects were an essential dietary component for our ancestors. Around the world today, some 1,500 arthropods, including cicadas and caterpillars, beetle and bees and spiders and scorpions, are regularly eaten by humans. From a Western perspective, eating insects seems completely alien and disturbing, and yet we consume a range of different foods, including several species of marine arthropods such as crabs, prawns and lobsters. I've often claimed that insects are no more than flying prawns, but people throw up their hands in horror and say that they would never eat insects. They say that insects are dirty and disgusting! Of course, neither of these things are true. The real reason why insects are not eaten in Europe and America is more to do with an ecological concept known as optimal foraging. Simply put, this is about how much energy you get back from the food you collect against how much energy you used up collecting it. If you are going to feed yourself and your family on insects in cool, temperate areas of the world, you are not going to do that well. In warmer parts of the world, where insects tend to be large or occur in huge numbers, eating them makes good sense.

In many Western countries, locusts, crickets and mealworms are routinely cultured and sold as live food for reptiles, tarantulas and other exotic pets, and they can be ordered from specialist suppliers online. You can also rear your own easily enough. Once, in Oxford, I cooked up some crickets for a large audience of children and, after a bit of encouragement, they were all eaten. Some children ate more than their fair share and the mother of one boy came marching towards me from the rear of the lecture theatre. 'My son has just eaten a handful of crickets!' she barked, nonplussed. I wasn't quite sure where she was going with this and asked her what the problem was. She looked at me as if she couldn't quite understand what had just happened and then said, 'At home, he doesn't even eat broccoli.'

Despite the recurring media interest, it has taken some time for insect-eating to be treated as something more than an oddity or a fad. Insect-eating is neither. In some parts of the Middle East, the price of meat drops when locusts are in plentiful supply. Elsewhere, dragonflies are caught and roasted on sticks. In Mexico, mealworm flour made from the roasted larvae of flour beetles is added to normal flour to increase the protein content of tortillas. If you want to give it a try, make some bread using a ratio of 1 gram of ground mealworm flour to 14 grams of wheat or maize flour. It is extremely tasty and nutritious. Even the celebrity chef Heston Blumenthal declared my mealworm bread to be delicious.

Times are slowly changing and the farming of certain insects as food for animals or humans is no longer seen as bizarre. Soon, industrial plants will be set up to use the larvae of the black soldier fly as both feed for livestock and a soil conditioner for recycled food waste. Now, you might not think that you eat insects, but it's logistically challenging to make food without a few fragments of insects cropping up. *The Food Defect Action Levels* handbook, published by the US Food and Drug Administration, lists the contamination levels at which action must be taken for a range of foodstuffs. For peanut butter, an American staple, it is an average of 30 or more insect fragments per 100 grams. For tomato juice, it is an average of 10 or more fruit fly eggs allowed per 100 grams. In my view, 10 or more times this level of 'defect' is highly unlikely to do you any harm. So, rather than avoid eating insects at all costs, the time may be right for a complete overhaul of our eating habits. The case for eating less meat is clear, but every human being, even if not that active, needs around 50 to 100 grams of protein a day. Farming insects might provide an alternative source. Still, I don't understand why we find it weird to eat insects, when there is an insect product that has long been a highly valued food.

Bee vomit and other useful things

Some European rock paintings, dating back at least 8,000 years, depict human figures collecting honey from wild bees, and honey has been mentioned in many early texts. There is no doubt that honey has provided humans and their ancestors with an important source of food for millions of years.

Honey is not just good to eat – it has long been used in traditional medicine to treat a number of minor illnesses such as burns and wounds, and was important in ceremonial and other rituals. But what actually is honey and how is it produced? Honey is a food made by bees for themselves and their developing larvae and is stored in wax cells. I enjoy asking schoolchildren where honey comes from – a common answer I get is that bees collect it from flowers. They're on the right track, but have missed out a critical step. Bees forage for nectar, a thin, sweet liquid produced by flowers as a means of attracting the bees. The bees drink the nectar and the flowers get their pollen picked up and transferred from flower to flower – the vital process of pollination. So far, so good. The nectar that the bees swallow is stored in the first part of the bees' gut – known as the crop, or honey stomach –and it may take many hundreds of flower visits to fill it. Enzymes produced in the gut of the foraging bee begin to act on the still-watery nectar, and when the forag-ing bee returns to the colony, it regurgitates the contents of its crop and shares it with many other bees. These bees then start to regurgitate and re-swallow the nectar over and over again. All the while, enzymes are breaking down the sucrose present in the nectar, converting it into a mixture of simpler sugars, fructose and glucose.

When it is finally regurgitated into storage cells, the liquid is still quite thin and must be thickened through evap-oration. The heat within the hive and constant wing-fanning by hundreds of worker bees gradually matures the honey to a point where it is thick enough and sweet enough that spoilage through fermentation will not take place. The cells are then capped and the contents kept until required or until a beekeeper takes the fruit of their labours from them. So, when you spread honey on your toast at breakfast what you are actually about to eat is bee vomit. The kids love that bit.

Today, nearly 2 million tonnes of honey are produced every year but, as humans have an endless capacity to debase anything good to make more money, some of it is watered down or adulterated with other sugars.

And what about the honeycomb? The hexagonal shape of bees' cells is no accident – it has evolved because it is the most economical shape and configuration that will hold the most honey for the least wax. Individual cells are built at an upward slope of about 13° to keep the honey inside from dribbling out. When honey bees start building a nest, they gather together in a ball, the inside of which soon heats up to 35°C, the temperature required for wax production. Wax is produced by eight glands on the underside of the worker bees' abdomen. Each gland secretes thin flakes of wax, which are then gathered and worked into shape with the front legs and the mandibles. Wild honey bees have to build their nest from scratch, but beekeepers provide their bees with a foundation sheet pressed out from pure beeswax, reinforced with wire and mounted in a frame. Beeswax has long been used in a great number of ways by humans as the basis for polishes, paints and as a sealer or lubricant – but mostly commonly to make candles that burn easily and cleanly.

Propolis, a sticky substance made by bees by mixing wax with the resin from various plants, is used by the bees as a gap filler and sealer, but it has been also used by makers of stringed instruments as the basis for varnishes. Some people believe that it is useful in the treatment of human ailments but there is no scientific evidence that supports these claims. Another bee product that some people claim has health benefits for humans is royal jelly. Royal jelly is a secretion produced by nurse honey bees to feed bee larvae that are destined to become queens. It is mostly water but also contains some proteins, simple sugars and fats, as well as a small amount of B vitamins, vitamin C and some other compounds that have an antimicrobial action. It is good for bee larvae but will not do anything miraculous for humans – no matter how much it costs.

Redcoats

Humans were dyeing textiles before the end of the Stone Age, and there is some evidence that it may date back even earlier than this. Dyes were initially obtained from natural materials such as mineral ores and plants, but a few thousand years ago, humans found that certain insects could help them stand out from the crowd.

The Cochineal Insect, *Dactylopius coccus*, has been used as the source of a red dye known as carmine since the second century CE. Being soft-bodied, sessile sap-suckers, the bugs have evolved to protect themselves from being eaten by producing a bitter-tasting defensive compound called carminic acid, and it is from this that the dye is made. Cochineal insects live on certain species of Mexican and South American cacti known as prickly pears, and the bright red fabric dye made from them was a valuable commodity in the Aztec and Mayan civilisations. The insects were harvested by simply brushing them off the cacti into baskets. They were then dried in the sun so that they could be stored without the risk of decay setting in. The dried insects were then exported or kept until the dye was to be made. Carmine was also used as a food colouring in the manufacture of cosmetics and in the pharmaceutical industry. Nowadays, carmine has been superseded by synthetic dyes, but the naturally produced dye (E120, if you're wondering) has had something of a revival in recent years and is still produced in some countries.

In my undergraduate years, I used carmine red for staining specimens for examination under a microscope, but did not know then where it came from. Neither did I know that the famous English army uniforms, the 'redcoats', owed their brilliant scarlet colouring to a scale insect. In the late eighteenth century, a senior British naval officer introduced prickly pear cacti and cochineal insects to penal colonies in Australia so that the British could have their own independent source of the approved scarlet dye. It didn't go well – all the insects died off and the cacti went on to form an impenetrable spiny no-go area about the size of New Zealand. The prickly pear was eventually brought to heel by one of the most successful biological control programmes ever. The saviour of the day was a small moth imported from South America. The larvae of the moth eat the cactus pads and, in the absence of any natural enemies of their own, were able to do exceedingly well. Of course, the moth has, by accident or through trade, ended up in other areas such as the southern United States, where it can threaten rare indigenous cacti. It seems that whenever we think we know what we're doing, things can turn out rather differently.

Insect spittle

The most luxurious thing you can wear next to your skin is a material made from the product of certain insects' salivary glands or, to put it plainly, insect spittle – also known as silk.

Silk is the name we use for the protein fibres spun by spiders and insects and used by them for capturing prey or protection. It is a remarkable material, and some current research is aimed at trying to reproduce its properties in artificial fibres. But what we most often refer to as silk is the natural fibre made by silk moths that is woven to produce the soft, lustrous textile of the same name. There are about 60 species of silk moths in the world, the best known of which is the Silkworm Moth, *Bombyx mori*. Silk cultivation originated in China more than 4,500 years ago, and it was so valuable that the Chinese strenuously guarded the secrets of its production. The secret did eventually leak out to neighbouring countries, where other species of silk moth were used. But it was not until 550 CE that the Chinese method was revealed, when two monks are believed to have offered up the secret to Eastern Roman Emperor Justinian I in exchange for favour. Once agreed, they smuggled some Silkworm Moth eggs, and the seeds of the white mulberry plant on which the caterpillars fed, to Rome, concealed in hollow canes. Had the monks been caught I imagine they would have come to a seriously sticky end.

As a result, Italy became an important silk producer and, gradually, the knowledge and technology spread to France and Spain. By the middle of the eighteenth century, there was even a small silk industry in England. The Silkworm Moth has been domesticated and selectively bred for so long now that it no longer survives in the wild. Silkworm Moth caterpillars are kept in large, well-ventilated rearing drawers, and they eat vast amounts of mulberry leaves in the course of their development. When fully grown, the caterpillars spin a cocoon, inside which they pupate. These are collected, boiled and the silk is wound off onto reels. A single cocoon may provide hundreds of yards of silk, and the threads of several cocoons are twisted together to make a single, strong silk thread. The silk from several thousand cocoons may be required to provide enough thread for a single dress. Nothing is wasted, as the boiled pupae are canned and sold as human food. I've tried them and can tell you that they really don't taste good at all, although it is entirely possible that I misunderstood the Chinese instructions on the tin.

And we don't just get things *directly* from insects – their relationships with other species, especially plants, have given rise to a whole cornucopia of chemical compounds that have become indispensable to humans.

A natural pharmacy

Animals that have been alive on Earth for as long as insects have evolved all manner of ways of staying safe. The venoms of ants and wasps aimed at vertebrates are well known, but insects have always had to protect themselves against an army of unseen microbial assailants. An insect larva living in water, soil, dung or rotting matter will be exposed, sometimes for long periods, to a huge number of pathogenic viruses and bacteria. Think of the decade or more that a cicada nymph spends underground, feeding on root sap.

But how do we find a chemical compound that might be useful as a pharmaceutical drug?

We could wait for a chance event, such as a stray fungal spore landing on bacteria growing on an agar plate, an occurrence that led to the discovery of penicillin. Or we can go 'bioprospecting' for new antibiotics. With the alarming rise of antibiotic resistance, this has become an urgent quest. Traditional or folk medicine has long used preparations derived from various insects, and the efficacy of a few of these is now being examined scientifically. Cytotoxic compounds have been found in insect venoms, and many other compounds known as peptides – short chains of amino acids – have been shown to have significant antimicrobial action. Fly larvae that live in decaying matter have powerful antibacterial peptides, and insects whose habits expose them to fungal infections produce agents that destroy fungal threads and even inhibit the germination of spores.

Anticoagulant compounds found in the saliva of blood-feeding insects hold the secrets of how we can make new clot-busting drugs. To date, most of the antibiotics we use have been found in soil, but the rate of discovery of new ones has slowed considerably in recent years. Perhaps it's time we took a look at insects and the biochemical riches they contain. We just need to understand how insects live, especially those that breed in biologically challenging environments, and ask ourselves how they survive. The drugs we swallow and inject in the next few decades may well have evolved inside the body of an insect.

We don't only take lessons from learning what compounds insects produce to protect themselves from harm – what's just as useful is what other organisms, such as plants, do to defend themselves from insects attacking them. There is a type of uniquely flavoured tea known as Eastern Beauty. It is believed that a Taiwanese tea-grower in the late nineteenth century ignored the fact that the buds and young leaves of his tea plants had been attacked by a type of sap-sucking bug known as a leafhopper. Rather than lose his entire crop, he took it to market anyway, where the tea made from the leaves had such a delicious and inter-esting flavour, reminiscent of ripe fruit and honey, that the grower was paid even more money than he would normally have expected. What brought about this happy reversal of fortune? The answer lies in the production of certain defen-sive compounds made by the tea plant in response to the attacks of the leafhopper.

The interaction between insects and plants is one of the most important relationships on Earth. For more than 350 million years, plants have been sucked and chewed by hordes of hungry insects. Ancient plants would have responded to being eaten by insects by evolving physical characteristics such as tougher leaves or spines, but these defences come at a cost: the more energy you use defending yourself, the less energy you have for growth and reproduction. Plants that were able to produce, as a by-product of their metabo-lism, chemicals that happened to repel or sicken herbivores would have become successful. Today, as result of this, plants contain a vast chemical arsenal. Because vertebrate animals have virtually the same biochemical pathways as insects, the defensive compounds of plants will affect us in similar ways. There is plenty of evidence that humans have made use of plant defensive compounds for thousands of years as poisons, medications, flavourings and cloth dyes without knowing what the active compounds were.

Over time, insects adapted their metabolism to defeat the plants' defences, and this evolutionary arms race has led to the production of more and more sophisticated compounds targeting insects' guts, nervous systems and other organs. Some compounds found in horseradish, cabbage and mustard taste good to us, but we have bred other crops to have fewer of these defensive chemicals in order to make them more palatable and, as a result, more prone to insect attack. But, happily, plants have provided us with natural pesticides too, such as pyrethrum from chrysanthemums and azadirachtin from the neem or Indian Lilac tree.

A multitude of compounds that plants evolved simply to keep insects at bay have improved our lives considerably and many of them, such as morphine, caffeine, nicotine, and quinine are familiar. The cancer drugs vincristine and vinblastine were derived from a flowering plant called Rosy Periwinkle, originally found in Madagascar, and taxol, an essential chemotherapy drug, was found in the bark of the Pacific Yew tree.

But we have only scratched the surface. There are thousands of plants whose chemicals and their properties we know nothing about. One person who knows a huge amount about the long and utterly fascinating relationships between plants and the insects that would eat them is Phil Stevenson.

PROFESSOR PHIL STEVENSON

The wonders of 400 million years of plant–insect warfare

At university I studied to be a zoologist and didn't have much interest in botany. I now realise how short-sighted I was. The relationship between insects and plants is the biggest terrestrial interaction on the planet. Knowing how they interact is where the real fascination lies. I interviewed Professor Phil Stevenson of Kew Gardens, and the first thing I asked him was how he would describe what he does.

'I study the chemicals in plants and look at how they mediate the interactions between plants and other organisms, especially insects. I guess you could call me a chemical ecologist.'

The two most important groups of organisms on Earth are plants and insects. The incredibly ancient interactions between them have produced some really interesting findings.

'In terms of the health benefits, ayurvedic medicine [the use of herbal remedies] goes back thousands of years. We may not have understood why these plants were doing what they were doing, but the story goes back a very long time.'

Nowadays, about a quarter of all the pharmacological drugs that we currently use are based on what plants create. Phil pointed out that if we include fungi in the important organisms list, then this increases to above 50% of drugs. I asked Phil why plants would go to the trouble of producing things that they don't necessarily need.

'Organisms produce these chemicals because they need to protect themselves. A plant can't run away when it's being attacked by a herbivore, so it develops various methods to protect itself. The methods that we're most familiar with include stings, prickles and cacti spines – but most commonly, plants use chemicals and they are all different.'

When land plants first appeared some 400 million years ago, they would have been fairly unprotected because they didn't realise that hordes of insects were about to evolve and start eating them.

'All of these mechanisms will have adapted through the need to survive. This process of evolution has generated an incredible array of natural products. In plants alone, there are between 50–100,000 different chemical structures from 350,000 specimens. When I talk about these chemicals, I'm referring to secondary metabolites. So, chemicals that are *not* absolutely essential for the plant's biochemical functioning or metabolism. Most people use some of these every day. The most well-known and well used, is the drug caffeine.

'Caffeine is produced by several groups of plants – not to get us moving in the morning, but to drive interactions with insects, particularly herbivores. We know caffeine is a defence chemical. That's why you get so much of it in a coffee bean – up to 4% of the weight of a coffee bean could be caffeine – it protects that bean from herbivory [being fed on]. Similarly with tea, you get very high concentrations of caffeine in the bud, the most precious leaf area, because it's the growing part of the plant. The plant produces these caffeine-filled buds because they're bitter-tasting, and are toxic to insects. However, a handful of insects have actually started to develop ways of tolerating that toxicity.

'It is an evolutionary arms race. The plants need to do a much more effective job at protecting themselves than the insects, because the insects can actually move off. Plants have to survive where they're rooted in the ground – otherwise, they will die. An insect's ability to adapt, through choices of different food sources, means they're less at risk than the plants themselves.

'The abundance of insects may also be a factor. There was a time when plants tolerated herbivory because they managed to grow faster than they were being eaten. The process of regenerating was effectively a defence mechanism. However, as the intensity of herbivory grew with the population of insects, those plants then had to start defending themselves with different mechanisms, whether it be spikes, spines or chemicals.

'However, if a plant focuses all its energy on not being eaten, it may not have enough left to produce as many seeds. It's a balance. Some plants have developed ways of producing defence mechanisms on demand. One species in the legume family, the chickpea plant (whose beans we use to make hummus), produces chemicals called phytoalexins if the plant has become infected, like an immune system protecting it from the invading organism.'

Many of our household drugs are the result of a long evolutionary history of predation between insects and plants.

'I would imagine most of the drugs in your cabinet protect against invading organisms. Salicylic acid is a compound produced by willow trees. If you acetylate this, you get aspirin. That would have been produced by willows as a defence mechanism against invading organisms. The most important analgesics in hospitals are opioids, from codeine to morphine and other related drugs. These are all plant-based opioids.

'One of the most fascinating areas is where we've developed anticancer drugs from plants. They wouldn't have been produced by the plant for their anticancer properties, but to protect against herbivory. Through the process of screening plant materials, we have actually found a biological activity against a specific cancer. One of the most well-known is perhaps the Madagascar periwinkle, a beautiful pink flower found in Africa, which is the source of two very complex alkaloids called vincristine and vinblastine. Before their discovery in the 1960s, if you were a child with leukaemia or non-Hodgkin's lymphoma, you would have had a 90% chance of dying. But now, you have a 90% chance of living. These are plant-based compounds that would have been produced by the plant to drive some ecological interaction, whether it was antifungal or anti-insect.

'We may never know about the usefulness of certain compounds in parts of the world that are now being destroyed. While the narrative around protecting rainforests has focused on maintaining the carbon capture potential of our forests, we need to consider the effects of deforestation on the potential medicine cabinets of the future. The chemical compounds found in rainforests are also much more diverse than they would be in temperate zones, so the need for these chemical defence mechanisms (and therefore potential usage) is much greater.'

It's not always the case that these plant chemicals have evolved to be exclusively used as defence mechanisms; some have multiple purposes.

'We could look at these processes as both antagonistic and mutualistic. In the antagonists' basket, we can put insects, fungi and bacteria. In the mutualists' corner, we have interactions with fungi, like mycorrhiza that enable trees to access important nutrients in the soil, or pollinators.

'One of the most interesting areas is where the same chemical is used for both protection *and* encouraging interactions. For example, if you give caffeine to most insects, it will taste bitter and act as a deterrent, or, in very high concentrations, even be toxic. However, if you decrease the concentration to quite low levels, something amazing happens. A bee, for example, will no longer find it distasteful, and will eat it quite happily. It has a pharmacological effect and improves the bee's memory for the traits associated with that flower. If a flower produces an attractive smell and contains caffeine in its nectar, that insect is better able to remember the smell the next day. When it goes out foraging again, it recognises the smell and goes straight to it because it knows it's food. As a result, it provides a more effective pollination service.

'What we don't know is if that was the *intended* function of caffeine, or whether it is simply meant to be a deterrent. Another amazing thing about caffeine is that it also has a medicinal effect on bees. It helps protect them against microsporidian parasites, which can be quite debilitating. These compounds have multiple functions in the ecological systems in which they exist.'

While we may benefit from some naturally occurring compounds, such as caffeine, there are a handful of plant substances that are rather addictive.

'This is a very interesting area. the compound THC (tetrahydrocannabinol) has now become part of our medicine cabinet because we've started to recognise its medicinal properties. The fact that humans have specific receptors for a compound produced by a plant and enables us to have a euphoric experience is remarkable.'

Nicotine is another plant compound known to be addictive, despite how unhealthy it is for humans.

'Nicotine is extremely toxic for insects. Despite being a fantastic insecticide, it is not advised for use in the garden by any means. It's the perfect example of a compound that was originally produced to protect the plant against insect herbivory that humans have found an alternative use for.

'It's another plant that we discovered by accident when looking for something to cure an upset stomach. I imagine it was originally eaten or chewed. Tobacco chewing still occurs in many parts of the world: in South America people still chew coca leaves; people in the Middle East chew khat, which has a similarly stimulant effect; and people in South Asia chew paan, which contains the betel leaf and the betel nut.

'I imagine, when we were originally experimenting with plants in order to cure ailments, we would have discovered all of their different effects.

'Of course, none of these compounds were produced by the plant to cure cancers, give us hallucinogenic experiences, or keep us awake. They were produced so that the plant could survive or, in some cases, as we're increasingly discovering, to harness the benefits of mutualists – the mycorrhiza or the pollinators.

'They've given us the most incredible chemical cabinet you could possibly imagine. If you think about how long humans have been trying to make different chemicals, we are nowhere near plants.'

Phil fears that our produce waste problem is doing nothing to help biodiversity or keep valuable species of plants from becoming extinct.

'The latest models indicate that 40% of plants on earth are currently at risk of extinction. This information has been generated through models based on our growing evidence of the distribution and diversity of plants in different locations. It is astonishing, and we still don't seem able to stop it.

'One of the reasons we need to preserve land is to produce more food, but the real challenge, I think, is making food more effectively from the land we already use for agriculture. There is an estimate that every year an area greater than the size of China produces food that is never eaten, because it's either lost or wasted.

'However, I do think there's been a turning point in the last few years. I feel optimistic, provided we can continue to support the science and the economics that helps us develop new ways of living. If we can cut how much we currently waste, that will massively ease the biodiversity crisis, which in turn helps slow the loss of plant species that may help with potential drugs in the future.'

I always knew it, but in talking to Phil, I realised the full extent of what insects have done for us. As an outcome of their constant war with plants, they have given us a hell of a lot of nice stuff to drink, eat and even cure ourselves with, and we should be incredibly grateful.

A written history

There are many things that distinguish humans from other animals, and perhaps one of the most important differences is that we are the only species to leave behind a recorded history. Whether or not we learn from it is an entirely different matter. We have scratched marks on sticks, left impressions in wet clay and carved symbols in rock, but a lot of history has been recorded using ink. Early forms of ink used soot, known as lampblack, but one type of ink has been in use for 1,400 years and is still used today; it comes from the complex interaction between a small wasp and an oak tree.

I live right next to Windsor Great Park, and here grows one of the biggest concentrations of veteran oak trees in the world. One tree standing in a private corner of the park is reckoned by some experts to be 1,300 years old. This means that the acorn from which it grew sent a little rootlet into the soil sometime during the eighth century CE. Incredibly, the tree is still alive and producing acorns. It is the longevity and endurance of the oak that makes it such an interesting species. I can walk past oak trees today that were growing in 1536, when Henry VIII rode to the top of a hill in the park to listen for the sound of a cannon being fired from the Tower of London. That distant bang confirmed to Henry that Anne Boleyn's head had just been lopped off.

The oak sits at the heart of a vast interconnected web of life, and there are more species of insects living on oak trees than on any other plant. Look closely at an oak tree and you will see all sorts of weird, tumour-like growths on the buds, leaves, twigs and even acorns. These growths, or galls, as they are known, are made by the oak tree in response to small gall wasps laying their eggs. Some, known as spangle galls, are small and abundant, while other less common galls are large and fleshy like little apples or hard like marbles. There are many different species, and the exact way each species of gall wasp manages to produce such an individual and unique gall is still somewhat of a mystery. Recent research suggests that the wasps are able to influence the early development of oak cells in highly specific ways – genetically engineering the oak tissues to grow a nursery and larder for the young gall wasp larvae inside.

But there is one type of oak gall that has shaped our history, because for more than a thousand years, it was the source of the ink with which nearly all our official documents were written. It is called the oak marble gall and the ink made from it is called iron gall ink. Crushed up, mixed with water, iron sulphate and gum arabic, the tannin-rich oak galls were transformed into a cheap and long-lasting ink. Some species of gall, such as one found on the Aleppo oak from the Middle East, had a far higher concentration of tannin than English marble galls and made better ink. For centuries, caravans of hundreds of camels carried sacks of these galls to be sold for the manufacture of dyes and ink. The last factory making dyes from oak galls, in what is now Iran, only closed a few decades ago.

But iron gall ink is not perfect. The ink is slightly acidic and this, over time, can cause paper and even parchment, made from animal skin, to disintegrate. Many documents survive with no damage at all, while others are badly affected and become completely eaten away. The reason for this is that the recipes used to make iron gall ink varied a great deal, both in the concentration of ingredients and the method of manufacture. I visited the National Archive at Kew, where well over a thousand years of British history – in the form of millions upon millions of documents written in iron gall ink – is stored. Here, I was able the read the documents written at the trial of Guy Fawkes, the best known of the Gunpowder Plot conspirators. He confessed after days of torture, but escaped the agonies of hanging, drawing and quartering by accidentally falling from the scaffold and breaking his neck.

Iron gall ink was used by American presidents and European governments for signing important documents well into second half of the twentieth century. This convenient quirk of evolution has brought us the Magna Carta, the works of Shakespeare and the American Declaration of Independence. We can read the only surviving copy of *Beowulf*, one of the earliest poems in English, and we can read the unique manuscript of the medieval romance *Sir Gawain and the Green Knight*. Thanks to iron gall ink, we can play the music of Mozart, Bach and Handel, examine the drawings of Rembrandt and Leonardo da Vinci, and read the correspondence of Charles Darwin. But how are we going to keep our history safe? The race is on to digitise a mountain of important historical documents and save them for posterity – but there are new worries, about the permanence of electronic storage. However, nothing – not even our planet – will last forever.

Model organisms

One of the most important ways in which insects are useful to us is as animals for scientific research. Much of what we know about biology, from genetics and physiology to behaviour and ecology, is based on studying the lives of insects. At school I dissected a locust and a cockroach as well as a rat and a dogfish, but many students today are reluctant to pick up a scalpel. I worry about how well they are going to fare without a first-hand knowledge of basic animal anatomy. I met a few of them when I taught at Oxford, and found it hard to understand – biology students who didn't want to cut anything up? 'We can read books,' they insisted, but their arguments are flawed. There is absolutely no substitute for doing dissections yourself. If we all took their view, we would still be stuck in the past with a medieval understanding based on inaccurate illustrations and defective descriptions.

Our understanding of life on Earth changed dramatically with the discovery of the structure of DNA – the molecule of life itself. In the short time since then, basically the time I've been breathing, the discovery has revolutionised biomedical science.

As chance would have it, I have not written a single word for the last hour because I went to make myself a coffee and was distracted by a number of small flies that had settled on some windfall apples I had collected – my primary teacher was right all along, a fly going past would indeed distract me from what I was doing! These are the flies that appear as if by magic in fruit bowls, the contents of which have long passed their 'best before' date. They belong to a family of more than

4,000 species, collectively known as fruit flies or vinegar flies. I'm pretty sure that the one I was looking at was the cosmopolitan species, *Drosophila melanogaster*. It is less than 3 millimetres long, with a yellowish brown body and distinctive bright red eyes. I tried to take a close-up photograph of one but had to be careful, as the fine hairs sticking out from its surface are sensitive to air currents, causing them to take flight at the slightest movement. At a glance you'd think it was an unremarkable little fly – a domestic inconvenience – but *Drosophila melanogaster* is arguably the most important insect species in the history of our world.

DNA contains all the information that allows living organisms to function and reproduce themselves, and although you might think that we and fruit flies are vastly different animals, you'd be wrong. Roughly 60% of the genes present in a fruit fly exist in humans. It seems animals are animals and animal cells are animal cells – they have a common ancestry after all, and work pretty much in the same way.

It was more than 100 years ago that the fruit fly flew from the fruit bowl to the scientific laboratory, where it now holds a preeminent position as a model organism. At the beginning of the twentieth century, biologists had been looking for an animal that could be used to study the nature of heredity in place of mammals such as rabbits, rats or mice. The work was expensive and time-consuming, and inbreeding quickly led to the appearance of damaging traits. Researchers realised that cultures of *Drosophila* flies could be inbred for many generations without any loss of health. They could be bred cheaply and easily and did not take up a huge amount of space. Moreover, with a generation time of less than 2 weeks, 25 generations could be produced in just a year. This greatly accelerated the amount of work that could be done. Chemicals or radiation were used to produce mutations in the flies, and it was not long before many mutations were discovered and their position on the chromosomes mapped.

To make life even easier, fruit flies only have 4 pairs of chromosomes, compared to the 20 pairs in mice. By the year 2000, the complete *Drosophila* genome, comprising some 15,000 genes, had been decoded. Today, we have all manner of molecular tools to switch genes on or off or alter them. Because fruit flies possess nearly all the genes that can cause cancer in humans, the study of cancer in *Drosophila* has enabled us to work out the complex biochemical pathways inside cells. These tiny flies are proving invaluable in the study of human conditions such as Parkinson's disease, Alzheimer's and muscular dystrophy, as well as revealing the underlying mechanisms of immunity, diabetes and ageing.

Picture the scene: it is a summer evening somewhere on the Mediterranean coast and a couple are enjoying a romantic supper al fresco. Although they are well past middle age, the light from the candles on the table flatters them. She is wearing a red merino shawl over her silk dress. He is wearing a crisp linen suit. Crickets chirruping softly in the undergrowth and flashes of fireflies in the bushes around them provide a natural son et lumière. They will enjoy a delicious meal of pan-fried trout served with courgette and tomato. For dessert, he will order figs with honey, while she will pick whatever she wants from a bowl filled with pears, nectarines and apricots. There will be coffee and tea to drink, with macadamia nuts and chocolate. Of all the nights they have had to remember in their lives, this will be the best. She has just learned that her cancer is well and truly on the run. She watches and smiles as her partner reaches over to take her hand – his tremors, once uncontrollable, have become almost imperceptible.

What have insects ever done for us, indeed!

Chapter 8

REPAIRING A WORLD OF WOUNDS

The beginning of the end?

There have been a few times in my life when I have visited places that are untouched by humans. I once lay down to rest in an unexplored cave system and I'm pretty certain that I was the first person to touch the cold, damp walls that surrounded me – but it's nowhere I'd want to live. After a week underground, living in profound darkness, illuminated only by torchlight, the sight of the outside world was overwhelming. The green of the forest below the rocky ledge where I emerged was intense.

But places like this are few and far between and, in the case of the cave system, it's clear why. In the incredibly short time since our appearance, we have covered the entire globe, establishing colonies wherever it was possible to survive. We have a deep-seated need to see 'what's over there'. A dozen of us have walked on the surface of the Moon, and some have visited the abyssal plains of the oceans, perhaps the last frontier of earthly exploration. But it seems we are not content with simply learning about these unique places – we want to mine them, to extract whatever we can get.

Today, it seems we live apart from nature. The impact we have had on the Earth is clearly visible from space, and the natural world has taken a battering. Perhaps it all started to change with the development of agriculture. This may well have been the point in our history when nature became something to be feared and tamed. Over 10,000 years ago, farming as a way of life really got going, with the cultivation of crops such as rice, millet and maize and the domestication of animals such as pigs, cattle and sheep. As diets improved and humans multiplied, more and more land was commandeered. Religions encouraged the view that nature was ours for the taking – and we took it. In the Bible, in Genesis 1:28, we read, 'Be fruitful and multiply, and fill the earth, and subdue it, and have dominion over the fish of the sea and over the birds of the heavens and over every living thing that moves on the earth.' Well, we've certainly done all of those things. The belief that human beings have the right to do as they damn well please has filtered down through the ages and is still with us. Of course, the words were written by us, and at the time, it probably seemed like a pretty reasonable way of going about things. That we can continue to treat the natural world as we see fit – that we can forever take without there being any consequences, is sheer folly. This 2,000-year-old way of thinking is ultimately doomed to failure because the natural world is the only reason we exist at all.

I sometimes wish I had been fascinated by something other than biology, as I seem to have been in a state of almost constant anxiety since I began my career. Back in the 1930s, the American ecologist and conservationist Aldo Leopold wrote, 'One of the penalties of an ecological education is that one lives alone in a world of wounds. Much of the damage inflicted on land is quite invisible to laymen.' Today, that damage is only too apparent and, despite it having been glaringly obvious for the last 50 years, we are only just beginning to realise that something needs to be done.

The overarching bad news is that biodiversity – the Earth's complement of living organisms and their genes – has declined and is continuing to decline at an alarming rate. It is estimated that currently there are well over 37,000 species threatened with extinction, including a quarter of mammal species and perhaps a fifth of all birds. Overall, around a third of all terrestrial vertebrates are experiencing considerable declines, which may ultimately lead to their extinction. Even three-quarters of all primates, the animal group to which we belong, are facing an extinction crisis. If we can't look after the 'furries and featheries', or even our own immediate biological family, what chance have insects got?

It seems that almost everything we do, from killing fleas on our pets with chemicals that end up in waterways, to the use of unnecessarily bright street lights, has substantial consequences for insect populations and, therefore, ecosystems. But it needn't be like this. Crucially, we need to be able to change how we do things when solid evidence reveals the error of our ways. It seems that old habits do indeed die hard. If we can't change, insects are in trouble and so are we. Why have we got ourselves into this mess? What caused it? And where are we going from here? Global warming is increasingly making it easy to imagine a world without humans. But what about insects? What about life more generally? If, as I hope this book has shown you, insects are the most incredibly adaptable, resilient and hardy creatures on the planet, might they be better placed to weather the changes we humans have brought about? Or might there be a time when insect life, along with *all* life on planet Earth, becomes extinct?

Where have all the insects gone?

Many lines of evidence suggest that there has been a massive decline in the abundance of insects in my lifetime. It should have been difficult to ignore, but it seems we have been doing just that. There are reasons why insects are so often over-looked. They are small, and leaving aside butterflies and bees, which are universally loved for their beauty and usefulness, insects are associated in the popular imagination with pestilence and famine. A question I have been asked a fair few times when speaking up for insects is 'So what if a species of insect I've never seen or even heard of becomes extinct – why should I care?' There is no suitably concise answer, because biodiversity and ecology are immensely complex and we have only just begun to understand how life on Earth works in any case. As John Muir, the pioneering Scottish ecologist, wrote in 1911, 'When we try to pick out anything by itself, we find it hitched to everything else in the Universe.'

When I was a boy, the front of the family car would soon become encrusted with dead insects, and most garages would stock products to ease the task of removing their dried remains from the paintwork and headlights. Ask anyone my age or older and they will tell you exactly the same story. Since leaving Scotland in 1975, I have lived in the south of England, where it is generally warmer – and yet, during June, July and August, the fronts of cars seem remarkably free of splattered insects. Something hugely significant has changed, but it took place so gradually that no one realised it was happening. This effect is known as the shifting baseline syndrome. When things are abundant, no one can imagine a time when this might not be the case. Even the passenger pigeon, once thought to be the most numerous bird on Earth, disappeared without any difficulty at all, and so too have many other species. If something common disappeared overnight, or next spring arrived and there were no bees, people would sit up and pay attention. But a slow, steady reduction will not ring any alarm bells. The numbers could drop by half or more, but as long as it happens gradually, it is easily missed. Year by year the baseline shifts to the left and becomes the 'new normal'. This phenomenon is universal.

Some studies indicate that more than 40% of all insect species are threatened with extinction, and that the rate of extinction is perhaps as much as eight times faster than that of mammals, birds and reptiles. But are insects in general heading for extinction, as some dire warnings in the media have intimated? 'Ecological Armageddon after dramatic plunge in insect numbers' and 'Hyper-alarming study shows massive insect loss', declared some recent headlines. I am pleased that this issue has been brought to the front pages, but big claims need big data to back them up, and it wasn't long before some of the research was under the spotlight. It was, after all, hard to believe.

While it could be argued that the research on which those headlines were based had some methodological flaws, the overall picture of what is happening to insect populations around the world is a gloomy one, and I have never read a single piece of work that suggests otherwise. That insects are much less numerous today than they were 50 or 100 years ago is no longer in any doubt – but could they ever become completely extinct?

Mass extinctions

There have been major episodes in the Earth's history when many species present at the time became extinct relatively quickly. These are known as mass extinction events. One of the biggest happened 250 million years ago at the end of the Permian, the long geological period that saw the appearance of reptiles with mammal-like features – our ancestors. Popularly known as the Great Dying, the Permian extinction saw the disappearance of a substantial fraction of the Earth's species. The causes are not fully understood, but involved acidification of the oceans together with a drop in oxygen levels and an increase in atmospheric carbon dioxide associated with substantial volcanic events. The biological recovery from this cataclysm took many millions of years. It is now thought that insects, which had appeared and diversified well before the Permian, were not nearly as badly affected by the mass extinction as large species were;

and although many insects did become extinct, many others appeared for the first time. It's no surprise that insects do much better during times of environmental crisis than large animals – they are natural-born survivors because they are small, require less food and habitat to survive, and breed extremely quickly.

The Age of Humans

There is a general consensus among biologists that the Earth is now in the middle of a mass extinction event, one that has been caused entirely by humans. Our activities have brought about such significant changes that, after due consideration, the part of the Earth's history that involves us and the changes we wrought should be named the Anthropocene – the Age of Humans. There are discussions as to what might represent a recognisable and consistent marker for the beginning of the Age of Humans – the emergence of agriculture, perhaps, or the increase in levels of radioactivity due to nuclear technology, or the sudden appearance of large quantities of plastic polymers. Whatever is decided, I think the Anthropocene will turn out to be represented by quite a thin stripe in the geological record.

The most significant cause of the dramatic decline of insects and most other terrestrial species, is quite simply, the loss, degradation and fragmentation of natural habitats – wilderness that we have taken over for own purposes. When I started my zoological training nearly 50 years ago, I never thought the world would have lost as much biodiversity as it has, nor that we would be asking the sorts of questions we are today. Questions like: how many species can we afford to lose? Are some species more valuable than others? Which species must we keep at all costs? It sounds like we've already thrown in the towel – already accepted that nothing much can be done and we must resign ourselves to surviving in an impoverished world. Some people have suggested that priority should be given to species that are evolutionarily distinct and globally endangered; these oddities on the tree of life have few, if any, close relatives – they are literally one of a kind. For example, the aardvark is the only surviving member of an entire order of mammals, and if it becomes extinct, we lose something more valuable than if we lost one more species of mouse or deer. Like the process of triage in an emergency ward, we need to decide the order of treatment of a large number of casualties, and it's unlikely that many insects will be high on the priority treatment list.

ALISON STEADMAN

A love of insects, revealed

Discovering nature at a young age is crucial to creating adults that are compassionate towards the natural world, whatever career path they embark on. I was delighted to chat to Alison Steadman, beloved British actress and ambassador for London Wildlife Trust, about her love of the natural world.

Though I have watched Alison on television many times, I did not know until recently that she has a great love for small creatures. I wanted to find out where her love for insects came from.

'I grew up in the suburbs of Liverpool. My whole family were into gardening, especially my dad, so we always had a beautiful garden. My favourite thing to do was to crawl around it. My dad grew rhubarb, and there were always caterpillars on it, and so I would spend time spotting the caterpillars.

'I remember I came across a caterpillar with brown fur, and I was fascinated by it. At the time, we had a wonderful guy who was mending our garden fence, so I showed it to him. He said to me, "In Africa, those things are the size of scarves. They wear them as scarves." I remember thinking, "Really? Scarves!" And then he said, "See that rhubarb? They use it as umbrellas in Africa because it's so big. They just walk around with a rhubarb leaf over their heads." Those are really happy memories.

'The only time I was ever frightened of an insect as a child was when I got in the bath, and my mum gave me a loofah to wash myself. I put it in the water and an earwig came out from inside the loofah and started to float around. I didn't like that – they are quite scary. They were very frequent at the time but I never see them anymore.

'Earwigs are one of the most remarkable insects. They look after their young really well – the female keeps her eggs in a little clutch in the soil and she licks them every day to make sure they haven't got any fungal spores on them. Lots of people don't realise that they can fly, but under their front wings are these beautiful semi-circular hind wings, which are folded up about four times. It's a real palaver to pack them away.'

Now a critically acclaimed actress, I wondered if there was a time when Alison veered towards the sciences, or always stuck to the arts.

'I always loved acting; dressing up and pretending to be other people. However, I also enjoyed art at school. I also considered becoming a vet because I loved animals so much, but when you're young, you don't really know anything about the world. I have always had a love for animals though.'

Alison explained what she would do when she collected beetles, slugs and spiders and other small insects as a child.

'When I did collect them, I would put them in a jar or a box and keep them in the shed. I wouldn't bring them into the house. I had a childhood that was full of wildlife.'

I asked Alison if her parents or anybody else in her family were interested in insects, which might have encouraged her fascination.

'Not really. I had an Aunt Hilda, who had a big fish tank and she would have snails in there to clean the tank. She explained to me that the water snails enjoyed crawling up the glass and cleaning the tank. I would often help her with injured birds that she'd taken in. There were lots of things going on at her house, but I wouldn't say that my parents were particularly interested in bugs.'

Alison explained what made her pursue the role of ambassador for the London Wildlife Trust.

'I think they do absolutely amazing work. I try to visit nature reserves whenever I can. When they approached me, I thought, "Gosh, it's a small way of helping such a wonderful group, and if it helps them, then yes, I'm going to do it."'

She has also written a lovely book for children, called Spider! *(Hodder, 2017), which tells the tale of a young lad who hates spiders. However, he realises that they're actually not so bad after all. I asked Alison what encouraged her to write the book.*

'Eighty per cent of the people I meet, whether they're children or adults, will say, "I hate spiders. I can't stand them," and I've always found that very odd. I can understand it if you live in Australia, where some spiders can be rather nasty and do awful things, but in this country, there's really very few spiders that might give you a bite, and they certainly won't kill you.'

In fact, in the UK there are only 10 or 12 species that have jaws big enough or strong enough to actually break your skin, and you would have to handle them quite badly to make them bite you – but it's an incredibly common fear. With the exception of wasps, which if you go and annoy them, will quite rightly sting you, we live in a pretty benign part of the world.

'I'm thrilled every time I find out a new fact about a creature. Once you start to read up about this ordinary little thing that you see in your bathroom you find out the most incredible things.

'One particular day when I was filming, one of the other actresses brought her son to the set for his seventh birthday. I was hanging about waiting for my scene, and this little kid was there on his own. He looked a bit shy and I thought, *What can I talk to him about?* So I said to him, "Hey, do you like spiders?" He replied, "No, I hate them. There was one by our television the other day and my dad tried to kill it, but it ran off." And I said, "Oh, I hope you didn't kill it." I'm not an expert, but I do know a few facts about spiders. I started to tell him, asking "Did you know you can tell if it's a boy or a girl? If it looks like it's got boxing gloves, it's a boy, and they shed their coats."

'The next day, his mum said to me, "Oh, my goodness, you've had such an effect on Raphael. He said to me, 'You mustn't kill spiders, Mummy – you must respect them,' and 'Did you know this? And did you know that?'" This child hadn't said anything to me at all during our interaction, but he'd taken it all in, and I was thrilled about it. I thought, *Gosh, if I can do that with one conversation, I'm going to try and write a book and see if I can influence kids and get them to not be so scared.*

'I think children pick up the fear of spiders from their parents. I've seen it a lot in the museum I used to work at. We had displays with live spiders in cages, and I would hear people going past and the mums saying, "Eurgh, don't touch that, don't look at it." And I thought, it's behind a piece of glass, it can't hurt you. Even if you held it in your hand, it wouldn't hurt you.

'If people could just look at spiders and learn about them, eventually they will begin to care about them. Every time I speak to a child, and every time they see a programme where spiders are presented positively, we are chipping away at this ridiculous view that anything with more than four legs is inherently dangerous.'

I asked Alison if anybody has contacted her to say that the book had transformed their view of insects.

'Since it was published, I've had some really good responses. A lot of people, including adults, have said, "When I read it, it really helped me," so that's great.'

Many people seem to think that we can exist on our own – like we don't need the natural world, or it's somehow external to us; they hold the view that we're not part of it. I asked Alison if she believes that attitude is the root of the problems that we now face.

'I hope that children and schools are becoming more educated about our planet and wildlife, and that we can't live without one another. When I was at school, we were taught quite a bit – I remember going on nature walks. When I got to my grammar school, that was one of the first things they did. They told us to look at the trees and their different shapes. I hope that that is now coming back.'

I feel really sorry for kids today, who don't seem to have the sort of freedom that we had as kids. Alison agrees.

'There was a recreation field nearby when I was growing up and that was my favourite place. And the birdsong in our garden was just great. My sister, who's 12 years older than me, lives just a little way out in South Liverpool, and the birdsong in her garden was also incredible. Now, all these years later, she rarely gets a bird in her garden. She gets pigeons, the odd magpie and that's it. It has completely changed.'

Sixty years on from when I was growing up in Edinburgh – when you went on a journey, the front of your car would be plastered with insects. Nowadays, I could drive around the whole summer in Oxfordshire and get nothing.

'Oh, yes. I remember my brother-in-law's family lived in Great Yarmouth. They would come over for holidays a couple of times a year and their car would be thick with insects on the windscreen. Now there's nothing.'

That's simply because of habitat loss and the huge amounts of pesticides that we use to produce cheap food. It's had a colossal effect in a short amount of time.

'I love going to garden centres and pottering around looking at all the plants – it's lovely – until you go into the actual shop, where there are products to kill anything you want – ivy, rats, pests. I know it's difficult to manage these things, but if you poison a snail or a slug and a hedgehog eats it, then the hedgehog gets poisoned and so it goes on. We've got to find better ways of dealing with this, and not just happily throwing poison down at everything.'

Alison then went on to explain how she manages her own garden.

'I live in a flat in London. There are eight flats in two Victorian houses, and we have a lovely, shared garden in the back. We're right next to a small woodland too so, when I put out birdseeds, we get lots of birds and I'm thrilled to bits. I don't think I could live without being able to look out of a window and see birds coming to the feeders.'

It sounds like Alison is quite the bird-spotter, so I asked her to elaborate on this.

'I'm no expert, but I do know all my garden birds. When the RSPB [Royal Society for the Protection of Birds] did their bird watch last year, me and my partner spotted 15 different varieties coming to the feeders – which, for living in a city, is pretty good. At one point, a goldfinch came and Michael jumped up, waving his arms and shouting "There's a goldfinch!" I said, "Stop it – you'll frighten them all away!" We were laughing our heads off.

'Watching the birds on the feeders is so nice because you aren't tempted to go make a cup of tea or look at this or look at that. You just concentrate on watching those birds. One day, Longtail tits suddenly arrived, half a dozen of them – we hadn't seen those for about six months. Again, we were absolutely delighted. I was putting my food out for the birds the other day, and we'd been away, so there'd been no food for a week. I'd only just opened the bag when a robin was at my feet looking up, going, "Where've you been, for goodness' sake? I'm starving – come on!"'

That sense of being in the moment – when you're looking at wildlife really intently and nothing else around you is important. No matter how depressed or unhappy I feel about something, after a couple of miles' walk in a woodland, I come home and I feel better. Alison agreed that in the last couple of years with lockdowns, a lot of people have reconnected with the natural world.

'I hope that people stay connected to it and don't revert back. One of the things I love is my small room that looks out the back, on to a terrace and then the garden. I learn my lines there. I've got a desk and my partner has the same. We sit there working and sometimes learning lines gets so exhausting (particularly now I'm getting older) that I think, *I'm fed up with this, that line won't go in* or, *Oh, how much more have I got?* At that point, I just say, *Stop.* I take my binoculars and go watch the birds for 20 minutes. I refresh and calm down before I go back to my desk. For me, it's heaven.'

I asked Alison the most important question of all: what her favourite insect was.

'I have so many that I like, but I love ladybirds because I think they're so pretty and they all have different numbers of spots. I think they're beautiful creatures.'

Nature-depleted

Compared to all other countries in the world, where would you place the United Kingdom in terms of how much of its biodiversity survives intact? In the top 10% perhaps? Think again. In the top 25%? No. Certainly in the top 50%, surely? No, I'm afraid not. For those of you who think the United Kingdom is still a green and pleasant land, you'd better sit down. In a recent analysis, the UK was revealed to be in the bottom 12%. The UK is in fact one of the most nature-depleted countries in the world. We are at the bottom of the class or, put another way, we are world leaders in how *not* to look after the natural world. We have lost much of our forests. We have lost nearly all of our wildflower meadows. We are still burning moorland for the sport of shooting birds. We are still extracting peat that should be left exactly where it is, locking carbon away, and we are encroaching year by year on whatever natural habitat still survives.

Animals that were once commonplace a couple of generations ago are now rare. In the face of intensive farming practices and the growth of transport, consumer demands and urban development, wild places and the species that live in them have suffered greatly. Many people have been making a loud noise about this appalling state of affairs and, as I near writing the end of this final chapter, the UK Environment Bill passed into law. The legislation aims to improve our natural environment and halt the decline of species by the year 2030, but we are really going to need to get our skates on to achieve anything significant in less than a decade. If we look after natural habitats and let areas that have been lost regenerate, insects and other animal life will thrive. If we do not, insect numbers will continue to dwindle, and the effects will be felt in every part of every food web – causing the ruination of once resilient communities of species.

Up the garden path

For many people a garden of some sort is the first bit of habitat they will see. I carried out my first biological project at the age of about 12. I wanted to map our small garden in Edinburgh and discover what sorts of animals it contained. Although I managed to identify most of the plants, when it came to the animals, I was soon overwhelmed by the scale of the task. In the end, I wrote up a sketchy account, which I still have in my bookcase. Much later, a naturalist called Jennifer Owen did do a proper job of surveying a suburban garden in Leicestershire. It took 35 years to compile a list of 2,673 species, which included 474 species of plants and 64 species of vertebrate, most of which were birds. But 80% of the species found in the garden – 2,135, to be precise – were insects and other invertebrates.

The idea of a garden has its roots in the view that we must somehow make order and beauty out of chaos. The roots of the word 'garden' mean some sort of enclosure, but really they are exclosures. These creations of our imagination had to be protected against the ever-present danger that they might revert to nature. Fences and walls were required to keep out grazing animals, and the eradication, by whatever means possible, of what we consider to be 'pests' and 'weeds' became a large part of a gardener's life and work.

Today, we are seeing the rise of the non-natural garden, the worst of which feature expanses of artificial grass. Many years ago I tried to work out how many small creatures lived on the front lawn of the Oxford University Museum of Natural History, where I worked. I used a vacuum sampler to suck up material from a known, fixed area of the grass into a fine mesh bag, taking care to collect lots of samples from many randomly selected parts of the lawn. The next job was to transfer my samples to small pots of ethanol and peer down a stereo-binocular microscope at them for hours on end in order to count and identify what I had caught.

By my reckoning, the lawn was home to a conservative figure of 10 million invertebrates – mostly tiny springtails, flies and other small insects. Just imagine how many more creatures there would be among the long grass of a lawn left to grow wild. As I worked, I remembered a series of cartoons in an edition of *Punch* magazine I saw when I was a child. It featured a family of the future. In one cartoon the father was vacuuming an artificial lawn. At the time I thought it was utterly ridiculous, and yet it has come to pass. You can even buy shampoo for your ersatz lawn and spray it with a synthetic, grass-like odour so that you can enjoy that freshly cut fragrance. I would rather sit under the spreading bough of an ancient oak tree or lie among wildflowers listening to the buzz and murmur of insects than walk through a manu-factured mockery of nature. A garden without insects is a thoroughly joyless and dispiriting place.

Dead wood

Growing up in Scotland, I never saw a European Stag Beetle, *Lucanus cervus*, until I was in my twenties and came down south. British beetles are generally small, retiring creatures who don't make much of a fuss, but stag beetles are large and most impressive. The males have hugely enlarged jaws which, just like the antlers of stags, are used to fight each other to gain access to females. The vast majority of their long life cycle, anything between three and seven years, is spent underground, where the larvae feed on decaying wood. In May and June, the adults emerge to mate, after which the females burrow down into the soil to lay their eggs. These beetles do not travel far, so if you find a female, it is more than likely she will have emerged from nearby. Stag beetles, surely a flagship species, have been becoming rarer all over Europe and in Britain, where they are restricted to the south and southeast of England. Their numbers have been declining steadily for the last 40 years. Much of their woodland habitat has been lost, and an obsession with tidiness has led to the wholesale removal of the tree stumps and dead wood on which they depend. The immense ecological value of dead wood is now being recognised, and by just leaving it where it is to rot naturally will help. It's not just stag beetles – leaving dead wood alone will ensure the survival of a third of the rarest insect species in Europe.

The preoccupation with neatness is also damaging many other insects, such as spring bees, whose food sources, including dandelions, are ruthlessly and regularly mowed all along our road verges every year. About half of England's hedgerows – vital refuges and wildlife corridors that act as linkages in our impoverished landscape – have been lost since the end of the Second World War. Hedgerows that remain are being damaged by agricultural spray drift or are mechanically flailed to within an inch of their lives.

If charity begins at home, then so too must conservation. The gardens of England cover more than five times the area of all our National Nature Reserves put together, so it's up to all of us to make sure that insects are looked after. It's not just big business that has brought us to where we are; we can all make a difference. We are the consumers of pesticides (in more ways than one). We are the buyers of the peat. We are the people who cut the stems of ivy clinging blamelessly to trees, thus depriving countless millions of insects of a critical source of autumn nectar.

Wholesale destruction

I have been lucky enough to spend some time in tropical rainforests, and have seen the biological riches they contain with my own eyes. Straddling the equator across three continents, these complex habitats hold more biodiversity than anywhere else on Earth. Much of it is unknown. It is hard to be sure what fraction of our planet's organisms live in rainforests, but the consensus is that it's over 50% and might be as high as 75%. This alone should make rainforests so important that the idea of losing them should be anathema – but, apparently, it's not. Reduced from around 12% of the total land surface area, they now cover under half that. Between August 2020 and July 2021, Brazil cleared an area of rainforest seven times larger than Greater London. They are being felled, burned and cleared for cattle ranching, timber, mining and growing cash crops.

One crop stands out. Humans have used palm oil for several thousand years, but the uses for this edible vegetable oil have multiplied so much that it can now be found in many processed foodstuffs, as well as beauty products, soaps and shampoos. World production of palm oil currently stands at around 75 million tonnes, but is expected to rise to more than three times this amount by 2050. Rainforests are being felled and burned wholesale and replaced with massive areas of oil palm trees to generate economic growth – to fuel cars, to make unhealthy food, to make face creams and lotions. Destroying rainforests to make money is, as the American entomologist Edward O. Wilson once wrote, 'like burning a Renaissance painting to cook a meal'. But any work of art, a man-made and relatively trivial thing, could be copied in every detail, using exactly the same materials. The contents of every art gallery in the world piled high are inconsequential compared to intact rainforest.

Most rainforest loss has taken place in the last 100 years, and it remains to be seen if we will come to our senses soon enough to save what's left. If rainforests are considered a vital resource of global importance, which they undoubtedly are, then they should be protected, and the countries that have them within their boundaries should be given adequate financial support to do exactly that. It's no good saying they're important if we, collectively, are not prepared to pay for them. We are all to blame.

In the future there will be people who might claim they didn't know what they were doing – that no one had told them such a disaster was unfolding. Well, it simply won't wash. Human beings started to slash and burn forests for subsistence in the late Stone Age and we have gone on slashing and burning ever since. Today, we can do it incredibly fast, and have reached the point where we could destroy all of the world's rainforests within the next century. If we do, we will lose at least half of the all the species on Earth. Showing exactly what we are in the process of losing, by highlighting the fate of insects, is the reason I left academia to become a television presenter.

Moth mayhem

As I carefully spread a white sheet over some low vegetation in a remote part of Papua New Guinea, I was vaguely aware that a helicopter was hovering a kilometre away. Although I could barely hear the thrum of the blades, I knew what I had to do as the director was talking to me on a walkie-talkie. 'OK George, when you've spread the sheet, stand back and look around you.' The scene was being filmed for the BBC expedition and subsequent TV programme *Lost Land of the Volcano* (2009). In the helicopter, an aerial camera operator was filming using a high-definition camera with a powerful zoom lens mounted on a stabilised, vibration-free mount. This sequence would reveal how tiny I was as I stood on the crater rim of Mount Bosavi, a volcano that last erupted over 200,000 years ago when early humans were leaving Africa, our birthplace, to populate the rest of the planet. The sequence would begin with me appearing full in the frame as I flattened out the white sheet, then, as I stepped back, the camera would pull back, gradually taking in a wider and wider view, with me shrinking until I would become a dot in the frame. The huge extent of the rainforest and the 4-kilometre width of the crater would be clear for all to see.

It promised to be a truly marvellous piece of television, but that night it started to rain; not just normal rainforest rain but biblical torrents – dam-busting amounts of rain. The plan was to attract moths to the white sheet using a high-powered ultraviolet lamp which had, amazingly, survived being hauled halfway around the world. I was convinced that the plan was completely doomed – surely no moths in their right minds would venture out on a night like this. The director, an unflappable professional of considerable experience, drew me aside and told me firmly that the moth-trapping was not going to be cancelled and reminded me that the next day we would have to make our way back to base camp in the valley below and that if we did not film the 'bloody moths' tonight, he had just wasted a considerable amount of money on a helicopter aerial shot that would be unusable. With a backwards look that was supposed to mean, 'Well, don't say I didn't warn you!', I went off to rig up some kind of rain cover for the lamp's bulb.

My mood had not improved by the time the light was switched on, but within 30 minutes, I was truly ecstatic. Against all the odds, the air around me was filled with moths – thousands of them – large ones with a wingspan the width of my hand, all the way down to tiny moths the size of rice grains. As they fluttered around, crawling on the sheet, my clothing and my face, it felt like someone off camera was chucking bucketloads of moths in my direction. I needed to explain this to camera. 'This shouldn't be happening,' I spluttered. 'Just look at this diversity – it's absolutely staggering!' And it truly was.

The important thing to remember is that the ultraviolet bulbs used for moth-trapping do not have that large a range – that is to say, most of the moths attracted to the light will have flown from less than 200 metres away. This means that all the moths that were flying that night had flown a relatively short distance for their 15 minutes of fame. To see that many species of moth flying – on the worst of nights, in a high-altitude mossy forest that clung to the rim of an extinct volcano in Papua New Guinea – was one of the most memorable filming experiences I have ever had. I went to bed late that night and the adrenaline rush kept me awake for an hour or so. The noise of the rain hammering on the tarpaulin roof was deafening and, as I drew my sleeping bag around me, I thought about all the things I could have said to camera.

The moth-trapping sequence on Mount Bosavi was a great hit with viewers and I have shown it many times in lectures. If the number of moths is anything to go by, then places like Mount Bosavi are special indeed – but they are also incredibly vulnerable. Animals and plants that live high on mountains are adapted to the cooler and wetter conditions found there, and many of them will not be found at lower altitudes. Because the area covered by the tops of mountains is much smaller than lower-lying areas, these habitats are exceedingly rare. The big threat to the species that live on top of the world is global warming. As temperatures rise and conditions get drier lower down, species will shift steadily upwards. But as they move up, they find themselves living in a smaller and smaller area. What happens to those species already living at the top? They have nowhere to go and will be lost forever. It is more than likely that a fair few of the moth species attracted to the bright light that night have never been described by science and that their fleeting appearance in our documentary might be the only evidence that they ever existed at all.

Poisoning the planet

Apart from extensive and substantial habitat loss, the other major cause of insect decline in recent years is the ever-increasing use of toxic chemicals used to produce cheaper and cheaper food an industrial scale. No sooner than we started growing large amounts of crops, insects, naturally, were on to it. Early farmers knew about different ways to combat pests, such as mixed planting and crop rotation. But they also knew about pesticides. The use of sulphur compounds to deter insect pests goes back at least 4,000 years. Pyrethrin, a naturally occurring insecticide obtained from the dried flowers of certain chrysanthemum species, has been used for at least 2,000 years and, like many modern insecticides, works by interfering with the way insect nerve cells work, resulting in paralysis and death. In time, crop plants were bred to have higher yields and be more nutritious. Whatever was good food for us was also good food for insects. And so began a chemical war on insects.

Eleanor Ormerod was a self-taught Victorian entomologist who, in her day, was lauded for her work on the control of agricultural pests, and hailed as the 'protectress of agriculture and the fruits of the earth'. Anything that ate our crops was the enemy. She recommended the mass slaughter of house sparrows and was an enthusiastic advocate of the use of Paris Green against orchard pests. Paris Green, a compound of copper and arsenic, was first made in the early nineteenth century as a superior green pigment for artists, but because it was extremely toxic, it was soon put to other uses – most notably in the control of the Colorado Potato Beetle that lived in the Rocky Mountains of North America, where it fed on a spiny nightshade species called Buffalo Bur before finding cultivated potatoes much more tasty.

After the Second World War, when food production was scaled up and mechanised, more synthetic insecticides were developed. At first people were unconcerned by what hazards these chemicals might pose but eventually it became apparent that they could affect our health and be dangerous to the environment. But there were other forces at work. During the twentieth century, insects became villains and were used as a metaphor, a proxy for all the evils and horrors that might befall us. Early horror and science-fiction movies often featured giant insects bent on world domination – even flesh-eating bees. It is likely that these films brought about a general acceptance that these alien-looking creatures should be feared and wiped out at any cost. Ridiculous, right? But

The Deadly Mantis, made in 1957, illustrates this mindset only too well. A 200-foot-long praying mantis locked inside polar ice for millions of years is suddenly freed thanks to a nearby volcanic eruption, and is naturally quite peckish. It goes on the rampage, devouring people right, left and centre, and generally making a nuisance of itself. Eventually, it reaches Washington, where, of course, it climbs the Washington monument to survey its new hunting grounds. The army and air force unleash their awesome firepower on the gigantic creature, to no avail, but science saves the day and the beast is despatched by means of insecticidal bombs – three cheers! Selling insecticides just became a whole lot easier.

A fatal flaw

One of the major problems at the heart of insecticide use is that they kill *all* insects. They do not discriminate between pest and non-pest insects, including beneficial ones such as ladybirds and bees. In the 1980s, a new class of insecticides, called neonicotinoids, was introduced. Like many insecticides, they work by irreversibly blocking the transmission of impulses in the insect's nervous system. These systemic nerve poisons, which were taken up by plants, were seen as a game changer because crops could be made toxic to insects all the time. Effective as insecticide is, a lot of it isn't absorbed by the plants and goes on to affect soil organisms, finding its way into streams and rivers, where it does yet more damage to aquatic insects. A survey of 16 English rivers carried out in 2017 found that half of them contained chronic or acute levels of neonicotinoid contamination. Treated plants have toxic pollen and nectar, and although the amounts might be small, they have enormous impacts on pollinating insects, especially bees. Even honey, a product that is often associated with purity, has been poisoned. A recent survey showed that three-quarters of honey samples from around the world contained measurable amounts of neonicotinoids – indeed, half of the samples analysed contained residues of a cocktail of two or more of these chemicals.

For all our efforts we are singularly failing to protect some of the most important insects of all. One of the worst practices is that, just like the excessive use of antibiotics in farm animals that are not sick, neonicotinoids are often used as an 'insurance policy' rather than in response to an actual insect attack. The dramatic increase in neonicotinoid use in the mid 2000s was driven almost entirely by its use as a seed treatment on a 'just in case' basis.

The agrochemical industry may say with confidence that, year on year, the weight of insecticides applied to crops has fallen quite a bit in the last 20 years. That might be so, but don't forget that many of the insecticides applied today are many hundreds if not thousands of times more toxic than anything that came before. There is a growing body of scientific evidence that the environment is now suffering under this constant chemical burden – farmland bird species continue to decline, bees and other pollinators are being poisoned. In my lifetime, insecticides and other pesticides have become so pervasive that they are in fresh water, in the food we eat and in the air we breathe. The flesh and organs of most humans today contain measurable residues of dozens of synthetic chemicals, the burden of which we know remarkably little. For example, in the United States, official studies have shown that the chemical signatures of DDT are present in 85% of all people tested, even though it hasn't been widely used for 50 years. As there are no reliable statistics available, it's hard to know exactly how many people suffer directly from exposure to pesticides – especially in countries where safety regulations are lax or absent – but it's likely to be many hundreds of thousands,

A report presented to the UN Human Rights Council in 2017 was extremely critical of the global corporations that manufacture pesticides, accusing them of, among other things, unethical marketing and lobbying which they consider to have "obstructed reforms and paralysed global pesticide restrictions". The question is: do we really need all these chemicals to feed the world? The agrochemical industry says we do, but the World Health Organization has shown that 30% of all food grown is already wasted, most of it in developed countries. It is estimated that the amount of food wasted per person every year in Europe or North America is around 100 kilograms – 10 times more than in less developed countries. In any case, increased food production has not, it seems, succeeded in eliminating hunger worldwide. Today, as many people die from coronary heart disease, type 2 diabetes, stroke and some cancers as a result of eating too much as die of eating too little. It might also come as a bit of shock that food wastage is responsible for 6% of the total global greenhouse gas emissions, which is three times the emissions produced by global aviation.

But this story is part of a much wider one. You cannot worry about habitat degradation and species loss without talking about global warming.

A warming world

Before we became technologically and scientifically capable, we believed that, apart from the odd glitch here and there, the climate was pretty stable. We were wrong. Seen over a larger timescale, this is far from true. The climate has, at times, been much warmer and also much cooler than it is today. The Earth has experienced many ice ages, when large parts of our planet were significantly colder and drier and sea levels were much lower. The ice sheets of the last ice age, which reached their greatest extent between 20,000 and 25,000 years ago, eventually retreated around 10,000 years ago – and as the ice melted, sea levels rose once more. Humans spread from isolated areas to cover much of the Earth's surface, and the warming conditions saw the rise of agriculture and the earliest civilisations. When I was growing up, there was much serious talk about another ice age being just around the corner, but it now looks like the next cold snap might be postponed until a later date.

The discovery and use of coal and oil changed everything. The sudden (geologically speaking) burning of prodigious amounts of carbon that had been locked away for millions of years fuelled the Industrial Revolution, but it has also altered the Earth's atmosphere. If we had only used wood and had instead maintained a balance between what we took and what grew back, we might have avoided global warming, but we would not have become the technologically advanced species we are today. We have known the likely consequences of pumping increasing amounts of carbon dioxide into the atmosphere for well over 100 years.

Now, in the face of this mess, we are, like Mr Micawber in *David Copperfield*, hourly expecting something extraordinary to turn up to make it all go away. It seems likely that we will have to suffer the long-lasting consequences of not abandoning the use of fossil fuels completely. Serious damage has already been done, and we will have to brace ourselves for a greater frequency of extreme weather events such as heatwaves, wildfires and droughts, as well as typhoons, cyclones and floods. As more and more permanent ice melts, sea levels will rise, submerging low-lying land completely. But now, terrified that our convenient source of energy is no longer an option, we are clearing rainforests to plant monocultures of biofuel crops that are seen by some as 'green gold'. It is fool's gold. Having filled the atmosphere with an excess of carbon dioxide, we now fell the forests that might have offered us some hope of salvation. We will, as we always have, remodel our world, and the effects of our actions will define our future.

It is certain that our success has come at a great cost to the rest of the natural world. But one phrase that really annoys me and has been printed on countless banners, T-shirts and badges is 'Save the planet'. The planet does not need saving. It's been here for 4.5 billion years and it will survive perhaps the same number of years again. If by 'Save the planet', we mean 'Save life on the planet', again it doesn't make much sense. Life on Earth has proved itself to be rather resilient, even tenacious, surviving all manner of cataclysmic changes and upheavals over its long history. There's little doubt that our planet and life in general will go on with or without us. Insects will survive and feature largely in any future world. What is really meant by 'Save the planet' is 'Save ourselves' – and only we can do that.

Trying to get people to care about the natural world has been frustrating, and there will be many people after me who will say much the same thing. But what will it take before we realise that we face an enormous threat to our survival? This threat is bigger than anything we have ever had to face before and, because it's bigger than many of us can imagine, we do little but talk about it. In recent years there has been a never-ending stream of bad news as far as the environment goes, and it can be incredibly depressing. But we must not let it paralyse us into inaction. The stakes are too high to sit idly by and wait to see what might happen next.

Take deforestation as an example. We know that the world needs trees. Trees shade and cool the planet, they store vast quantities of carbon and regulate the global water cycle, as well as forming communities that comprise over half the world's biodiversity. But before we rush headlong into planting millions upon millions of trees to solve the problems we have made, let's consider our options. A plantation of a few fast-growing species is nothing like a natural woodland or forest. Sure, we may capture some carbon – but in our rush to do something quickly, we might be doing the wrong thing yet again. This is the time to let forests regenerate. We must stop cutting them down. Leave them alone and they will grow back faster than we think. Even regrowth or secondary tropical forests can regain many of their unique attributes in only a few decades and attain near original condition in less than 100 years.

Boom and bust

Should we be worried? Well, it all depends on your point of view. Ninety-nine per cent of all species that have ever lived on Earth are now extinct, and it was mass extinction events that paved the way for the appearance of those creatures that were the forebears of human beings. But compared to the insect life on Earth, we are mere johnny-come-latelies. There are days when I can't decide whether it would be better to just let the human species carry on the way it has been, or try to do things differently. We certainly know enough to do the things that might just prolong our own survival. One of the things we simply must do is appreciate and look after the natural world a lot more. To do that, we need to appreciate insects a whole lot more.

Some people say that there will come a time when we will have to leave the Earth to find a new home somewhere else. This idea has always been a popular science-fiction storyline because it allows us hope that no matter what we do here on Earth, we will somehow survive by going elsewhere. But if we do ever leave Earth, we will feel the need to take a number of plant and animal species with us to ensure a stable supply of food, raw materials and recycling services. We already know the folly of introducing alien species to places where they don't belong, but nevertheless, those brave pioneers would have to build themselves an ecological pyramid that would be self-sustaining and provide them with a life-support system we simply take for granted here and now. It's certainly something worth thinking about.

In Arizona between 1987 and 1991, Biosphere 2 transformed this thought into a reality. A facility composed of a number of linked, sealed habitats or biomes was constructed for eight people. The entire system was designed to function without the need for external input, except for sunlight. The idea was to see if a closed system to support human life in space was a viable proposition. It was an incredibly audacious project, which was technically and psychologically extremely challenging. It started off well enough, but after a while, oxygen levels began to fall, the seawater zone became too acidic and food ran short. Many of the insects that had been introduced at the start failed to thrive, while other generalist species such as cockroaches and ants began to run riot. To make matters even worse, the biospherians themselves split into two opposing factions and stopped working as a team. Does this remind you of anything at all?

Unlike bacteria, we are probably unsuited to being galactic pioneers – and for a species that cannot even deal efficiently with its own excrement, I do worry about the future. It would be much more sensible to put our own house in order here on Earth before we start to think about making a mess somewhere else. There may come a day when humans will venture far enough into space to visit other rocky worlds on which life has developed; however, the closest candidate identified to date is many trillions of miles away. I did think about this as I lay on the warm savannah looking up in sheer wonder at the starry African sky. I still find it hard to imagine that we will ever find another home. But never say never. If we do colonise other planets where life has evolved to produce multicellular organisms, I'll bet you anything they'll look a lot like insects. It's such a pity I won't be around to be proved right. When I die, I'd like to leave behind a *Space Odyssey*-style monolith made out of the hardest known material, engraved with the words 'I told you so' beneath an image of a beetle.

Repairing a world of wounds is not going to be easy but, as Jacques Cousteau, the French pioneer of marine conservation, wrote, 'If we go on the way we have, the fault is our greed and if we are not willing to change, we will disappear from the face of the globe, to be replaced by the insect.' Perhaps deposed is a better word – insects were here long before we appeared on the scene. Insects have already survived several enormous global catastrophes and will do so again.

Before I finish I'm going to leave you with a simple task. Find a bit of woodland, a grassy area or a hedgerow and sit down on the ground. Look at whatever small patch of habitat is right in front of you. Keep quite still and ignore everything else around you. Soon you will become aware of beetles and flies and many kinds of small insects crawling and flying through your field of view. Do not ever forget that insects lived on Earth long before even our primitive fish-like ancestors propped themselves up on fin-like limbs in the shallows to gaze out at dry land – and insects and their kind will still be here when we've gone. There's absolutely no doubt we're smart. The big question is: are we smart enough?

EPILOGUE

Our progress has been manifestly meteoric. It's only a little over 400 years since Galileo first gazed at the heavens with a primitive telescope to see the many hundreds of stars that could not be seen with the naked eye. I am glad to say that he wasn't always looking up, and that sometimes he used his pioneering optics to see insects more clearly. But even as I write this, the James Webb Space Telescope, a hugely complex and expensive piece of machinery, has just been launched successfully and is orbiting the Earth at a distance of about 1 million miles. Its instruments have allowed us to see light from the most distant stars and galaxies – light that has been travelling through space for billions of years. We are now able to see further back in time than ever before. But I can't help wondering what benefit it will be to us if we continue to exterminate life on Earth. We've come a long way since we believed that our planet was at the centre of the universe – perhaps it's time we also realised that we might not be such consequential creatures as we imagine.

Others will argue that bacteria or fungi are the most important organisms of all, and they may well be right. But

I have spent my whole life learning about insects, and to the day I hang up my hand lens, I will be their advocate. Insects are getting a bit more of a mention these days, but we need to realise that we owe them an enormous debt. Insects are excellent at adapting to environmental changes. Whatever our chances are, the chances of insects surviving far into the future are orders of magnitude higher than ours – they are future-proofed like no other animal, and that alone gives me great comfort.

ACKNOWLEDGEMENTS

My biological career has been guided and enriched by two very special people to whom I will always be immensely indebted. Dr Henry Bennet-Clark taught and mentored me at Edinburgh University during my undergraduate years, and Professor Sir Richard Southwood supervised my doctoral studies. In time I become a colleague of theirs at Oxford University – I was doubly fortunate.

I am most grateful to Sir David Attenborough, Professor Stephen Simpson, Professor Helen Roy, Professor Philip Stevenson, Professor Karim Vahed, Dr Erica McAlister and Alison Steadman for agreeing to be interviewed and for giving their time so generously. I must also thank Anne Riley of the Wharfedale Naturalists' Society for carefully reading the manuscript.

FURTHER READING

I have included a handful of books that will allow the reader to learn about the biology, classification and importance of insects in more depth.

Chapman, R.F., Simpson, S.J. and Douglas, A.E. (2013). *The Insects: Structure and Function* (5th edn.). Cambridge University Press.

Chittka, L. (2022). *The Mind of a Bee*. Princeton University Press.

Goulson, D. (2022). *Silent Earth: Averting the Insect Apocalypse*. Vintage.

Leather, S. (2022). *Insects: A Very Short Introduction*. Oxford University Press.

McGavin, G.C. and Davranoglou, L.R. (2022). *Essential Entomology* (2nd edn.). Oxford University Press.

Sumner, S. (2022). *Endless Forms: The Secret World of Wasps*. William Collins.